COMPLEX ANALYSIS

UNIT A1 COMPLEX NUMBERS

Prepared by the Course Team

Before working through this text, make sure that you have read the *Course Guide* for M337 Complex Analysis.

The Open University, Walton Hall, Milton Keynes, MK7 6AA.

First published 1993. Reprinted 1995, 1998, 2001, 2008

Edited, designed and typeset by the Open University using the Open University TEX System.

Printed in Malta by Gutenberg Press Limited.

ISBN 0 7492 2175 5

This text forms part of an Open University Third Level Course. If you would like a copy of *Studying with The Open University*, please write to the Central Enquiry Service, PO Box 200, The Open University, Walton Hall, Milton Keynes, MK7 6YZ. If you have not already enrolled on the Course and would like to buy this or other Open University material, please write to Open University Educational Enterprises Ltd, 12 Cofferidge Close, Stony Stratford, Milton Keynes, MK11 1BY, United Kingdom.

1.4

CONTENTS

INTRODUCTION

Historical survey

Before describing the contents of this unit, we provide a brief historical account of the development of complex numbers. Many mathematical problems that have been studied since ancient times lead to quadratic equations of the form

$$ax^2 + bx + c = 0, \tag{0.1}$$

where a, b, c are real numbers, with $a \neq 0$. The formula

$$x = \frac{-b \pm \sqrt{b^2 - 4ac}}{2a},$$

for the solutions, or roots, of Equation (0.1), gives:

   two distinct real solutions if $b^2 - 4ac > 0$;
   one (repeated) real solution if $b^2 - 4ac = 0$;
   no real solutions if $b^2 - 4ac < 0$.

When $b^2 - 4ac < 0$, Equation (0.1) has no real solutions because a negative number cannot have a real square root.

However, as long ago as the 16th century, the Italian mathematician Cardano began to experiment with the manipulation of symbols such as $\sqrt{-1}$, using the ordinary rules for real numbers. For example, he considered the problem of finding numbers x and y such that

$$x + y = 10 \quad \text{and} \quad xy = 40. \tag{0.2}$$

This problem has no real solutions, since it reduces to the quadratic equation

$$x^2 - 10x + 40 = 0,$$

In this case,
$$b^2 - 4ac = -60 < 0.$$

which has no real solutions. Cardano pointed out, however, that if $\sqrt{-15}$ is treated in the same way as a real number, then

$$x = 5 + \sqrt{-15} \quad \text{and} \quad y = 5 - \sqrt{-15}$$

do satisfy Equations (0.2). At the time there was little enthusiasm for such a solution because of doubts about the existence of entities such as $\sqrt{-15}$, but eventually Cardano's idea proved its worth in a rather unexpected way, which we now describe.

The general cubic equation

$$ax^3 + bx^2 + cx + d = 0, \tag{0.3}$$

where a, b, c, d are real numbers, with $a \neq 0$, is much more difficult to solve algebraically than is the quadratic equation. The following remarkable method was known to Cardano and other Italian mathematicians in the 16th century.

First, Equation (0.3) is reduced to the form

$$x^3 + px + q = 0, \tag{0.4}$$

There is no need to follow in detail the manipulations that lead to Equation (0.6), which provides a solution of Equation (0.4).

by substituting $x - \frac{1}{3}(b/a)$ in place of x and dividing through by a. Next, the substitution

$$x = u + v, \quad \text{where } uv = -\tfrac{1}{3}p,$$

is made (assuming that such real numbers exist). This transforms Equation (0.4) into

$$u^3 + v^3 + q = 0; \tag{0.5}$$

that is,

$$u^6 + qu^3 - (p/3)^3 = 0.$$

This is a quadratic equation in u^3 with solutions

$$u^3 = \frac{-q \pm \sqrt{q^2 + 4(p/3)^3}}{2}.$$

Now, by Equation (0.5),

$$v^3 = \frac{-q \mp \sqrt{q^2 + 4(p/3)^3}}{2},$$

and so a solution to Equation (0.4) is

$$x = \sqrt[3]{\frac{-q + \sqrt{q^2 + 4(p/3)^3}}{2}} + \sqrt[3]{\frac{-q - \sqrt{q^2 + 4(p/3)^3}}{2}}. \qquad (0.6)$$

Formula (0.6) works extremely well in some cases. For example, the equation

$$x^3 + 3x - 4 = 0$$

has $p = 3$ and $q = -4$, so that

$$\begin{aligned} x &= \sqrt[3]{2 + \sqrt{5}} + \sqrt[3]{2 - \sqrt{5}} \\ &= \tfrac{1}{2}(1 + \sqrt{5}) + \tfrac{1}{2}(1 - \sqrt{5}) = 1, \end{aligned}$$

which is indeed a solution.

You can check these cube roots by using the formula
$$(a + b)^3 = a^3 + 3a^2b + 3ab^2 + b^3.$$

However, for some values of p and q there is a difficulty. The equation

$$x^3 - 15x - 4 = 0 \qquad (0.7)$$

has $p = -15$ and $q = -4$, so that

$$x = \sqrt[3]{2 + \sqrt{-121}} + \sqrt[3]{2 - \sqrt{-121}}.$$

This solution involves the expression $\sqrt{-121}$, which suggests that Equation (0.7) has no solutions. However, following Cardano and treating $\sqrt{-1}$ in the same way as a real number, we obtain

$$(2 + \sqrt{-1})^3 = 2 + \sqrt{-121} \quad \text{and} \quad (2 - \sqrt{-1})^3 = 2 - \sqrt{-121},$$

so that

$$x = (2 + \sqrt{-1}) + (2 - \sqrt{-1}) = 4,$$

which is indeed a solution of Equation (0.7). Thus, by allowing the use of symbols which seemingly have no meaning, we can produce one correct solution to the original problem.

This method of solution did not lead to an immediate flowering in the study of such 'imaginary' numbers (as they were called), which continued to be regarded with great suspicion. A century later, for example, Newton stated that if the solution to a problem involved imaginary numbers then the problem had no *genuine* solutions. By the 18th century, however, mathematicians such as (Johann) Bernoulli, Leibniz and Euler were using such numbers in applications to integration, although there continued to be controversy about, for example, the definition of the logarithm of $\sqrt{-1}$.

The symbol i for $\sqrt{-1}$ was introduced by Euler, and the name *complex number* for an expression of the form $z = x + iy$, where x and y are real, was introduced by Gauss at the end of the 18th century. Gauss also made use of the geometric interpretation of a complex number $x + iy$ as a point with rectangular coordinates (x, y) in a plane (see Figure 0.1).

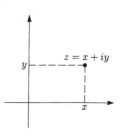

Figure 0.1

A major justification of the use of complex numbers came in 1799 when Gauss proved the so-called Fundamental Theorem of Algebra. This states that every polynomial equation

$$a_n z^n + a_{n-1} z^{n-1} + \cdots + a_1 z + a_0 = 0, \qquad (0.8)$$

where $a_0, a_1, \ldots, a_n$ are complex numbers, with $a_n \neq 0$, has at least one complex solution (and hence n complex solutions, some of which may be repeated). Thus, although it was necessary to introduce the new 'complex' numbers in order to solve quadratic equations, there was no need to introduce any further new numbers in order to solve all polynomial equations with complex coefficients. We should hasten to add that Gauss did not discover a formula for solving Equation (0.8) (in fact, in 1824 Abel proved that no such formula exists if $n \geq 5$), but he did demonstrate the *existence* of a solution.

Any lingering doubts about the validity of complex numbers were laid to rest in the 1830s when Hamilton gave a definition of complex numbers as ordered pairs of real numbers (a, b), subject to the rules of manipulation

(a, b) corresponds to $a + ib$.

$$(a, b) + (c, d) = (a + c, b + d),$$
$$(a, b) \times (c, d) = (ac - bd, ad + bc).$$

For example:
$$(1, 2) + (3, 4) = (4, 6),$$
$$(1, 2) \times (3, 4) = (-5, 10).$$

This had the effect of placing complex numbers on a sound algebraic (as well as geometric) basis.

The groundwork had now been laid for the great flowering of the study of complex numbers associated with the names of Cauchy, Laurent, Riemann and Weierstrass, which went on throughout the 19th century. Although most of the course will deal with the classical theory of complex numbers and complex functions, at the end of the course you will see some applications and also discover that, even towards the end of the 20th century, there are still new and exciting developments in the subject.

Introduction to this unit

The first block of the course is an introduction to complex functions, and it is designed to familiarize you with their most basic properties. This unit is devoted solely to complex numbers themselves.

In Section 1, we define complex numbers and show you how to manipulate them, stressing the similarities with the manipulation of real numbers.

Section 2 is devoted to the geometric representation of complex numbers. You will find that this is very useful in understanding the arithmetic properties introduced in Section 1.

In Section 3 we discuss methods of finding nth roots of complex numbers and the solutions of simple polynomial equations.

The final two sections deal with inequalities between real-valued expressions involving complex numbers. First we use inequalities in Section 4 (the audio-tape section) to describe various subsets of the complex plane. Then we show you, in Section 5, how to *prove* such inequalities. In particular, we introduce the Triangle Inequality, which can be used to obtain an estimate for the size of a given complex expression.

Study guide

After studying this unit you should be able to perform basic algebraic manipulations with complex numbers and understand their geometric interpretation. Before you tackle later units, it is important that you become confident with these basic manipulations, and we advise you to attempt as many of the problems and exercises (which follow Section 5) as you have time for.

Associated with this unit is a segment of the Video Tape for the course. Although this unit text is self-contained, access to the video tape will enhance your understanding. Suitable points at which to view the video tape are indicated by a symbol placed in the margin.

1 INTRODUCING COMPLEX NUMBERS

After working through this section, you should be able to:

(a) determine the *real part*, the *imaginary part* and the *complex conjugate* of a given *complex number*;

(b) perform addition, subtraction, multiplication and division of complex numbers;

(c) use the Binomial Theorem and the Geometric Series Identity to simplify complex expressions.

1.1 What is a complex number?

We assume that you are already familiar with various different types of numbers, such as the **natural numbers** $\mathbb{N} = \{1, 2, 3, \ldots\}$, the **integers** $\mathbb{Z} = \{\ldots, -2, -1, 0, 1, 2, \ldots\}$, the **rational numbers** (or fractions) $\mathbb{Q} = \{p/q : p \in \mathbb{Z}, q \in \mathbb{N}\}$, and the **real numbers** $\mathbb{R}$, which can be represented by decimals (terminating or non-terminating). We assume also that you are familiar with the usual arithmetic operations of addition, subtraction, multiplication and division of real numbers.

For example:
$\frac{1}{2} = 0.5$,
$\pi = 3.1415\ldots$.

We are now going to introduce the idea of a complex number and we begin with some definitions.

Definitions A **complex number** z is an expression of the form $x + iy$, where x and y are real numbers and i is a symbol with the property that $i^2 = -1$. We write

$$z = x + iy \quad \text{or, equivalently,} \quad z = x + yi,$$

and say that z is expressed in **Cartesian form**. The real number x is the **real part** of z (written $x = \operatorname{Re} z$) and the real number y is the **imaginary part** of z (written $y = \operatorname{Im} z$).

Two complex numbers are **equal** if their real parts are equal *and* their imaginary parts are equal.

The set of complex numbers is denoted by $\mathbb{C}$.

Here are some examples of complex numbers z which correspond to given real numbers x and y.

$z = x + iy$	$1 + 2i$	$\sqrt{2} + i\pi$	$3i$	1	$1 + i$	0	$1 - 2i$
$\operatorname{Re} z = x$	1	$\sqrt{2}$	0	1	1	0	1
$\operatorname{Im} z = y$	2	π	3	0	1	0	-2

Various conventions are commonly used with complex numbers, as you can see in the table:

(a) any real number x can be thought of as a complex number whose imaginary part is zero, and the complex number x is then said to be *purely real* (thus $\mathbb{R}$ is a subset of $\mathbb{C}$);

$1 + 0i = 1$

(b) if the real part of a complex number is 0, but the imaginary part is non-zero, then we write the complex number in *purely imaginary* form;

$0 + 3i = 3i$

(c) $0 + 0i$ is written 0, the zero complex number;

(d) we usually abbreviate $1i$ to i;

(e) if y is negative, then we usually write z as $x - |y|i$.

$1 + (-2)i = 1 - 2i$

1.2 Arithmetic in $\mathbb{C}$

The definition of a complex number contains the symbol '+' and refers to the 'square' of i. This suggests that arithmetic operations can be performed with complex numbers; the following definitions are made.

Definitions The binary operations of **addition**, **subtraction** and **multiplication** of complex numbers are denoted by the same symbols as for real numbers and are performed by the usual procedure — that is, treating the complex numbers as real expressions involving an algebraic symbol i with the property that $i^2 = -1$.

In some contexts, e.g. electrical engineering (where i is used for current), it is common practice to write j for i.

Some examples will help to make this definition clear.

Example 1.1

Express each of the following numbers in Cartesian form.

(a) $(1 + 2i) + \left(\frac{1}{2} + \pi i\right)$

(b) $(1 + 2i)\left(\frac{1}{2} + \pi i\right)$

(c) $2(1 + 2i) - 2i\left(\frac{1}{2} + \pi i\right)$

(d) $(1 + 2i)(1 - 2i)$

Solution

(a) By the usual procedure:

$$(1 + 2i) + \left(\tfrac{1}{2} + \pi i\right) = 1 + 2i + \tfrac{1}{2} + \pi i$$
$$= \tfrac{3}{2} + (2 + \pi)i.$$

(b) By the usual procedure:

$$(1 + 2i)\left(\tfrac{1}{2} + \pi i\right) = \tfrac{1}{2} + \pi i + i + 2\pi i^2;$$

applying the extra property that $i^2 = -1$, we obtain

$$(1 + 2i)\left(\tfrac{1}{2} + \pi i\right) = \left(\tfrac{1}{2} - 2\pi\right) + (\pi + 1)i.$$

(c) By the usual procedure and the property that $i^2 = -1$:

$$2(1 + 2i) - 2i(\tfrac{1}{2} + \pi i) = 2 + 4i - i - 2\pi i^2$$
$$= (2 + 2\pi) + 3i.$$

(d) By the usual procedure and the property that $i^2 = -1$:

$$(1 + 2i)(1 - 2i) = 1 - 2i + 2i - 4i^2$$
$$= 1 + 4 = 5. \quad \blacksquare$$

The following problems provide practice at such manipulation of complex numbers.

Problem 1.1

(a) Express each of the following in Cartesian form.

    (i) $(2 + i) + 3i(-1 + 3i)$

    (ii) $(2 + i)(-1 + 3i)$

    (iii) $(-1 + 3i)(-1 - 3i)$

(b) Write down the real and imaginary parts of $z = (2 + i) + 3i(-1 + 3i)$.

Problem 1.2

Express each of the following in Cartesian form.

(a) $(x_1 + iy_1) + (x_2 + iy_2)$ (b) $(x_1 + iy_1) - (x_2 + iy_2)$

(c) $(x_1 + iy_1)(x_2 + iy_2)$ (d) $(x + iy)(x - iy)$

If x and y, with or without subscripts, appear in the specification of a complex number, then it is presumed that x and y are both real.

As with real numbers, the *negative*, $-z$, of a complex number z is defined in such a way that $z + (-z) = 0$.

Definition   The **negative**, $-z$, of a complex number $z = x + iy$ is

$$-z = (-x) + i(-y),$$

usually written $-z = -x - iy$.

For example,

$$-(1 + i) = -1 - i.$$

Next, we discuss *division* of complex numbers. As with real numbers, the reciprocal, $1/z$, of a non-zero complex number z is defined in such a way that $z(1/z) = 1$.

Definitions   The **reciprocal**, $1/z$, of a non-zero complex number $z = x + iy$ is

$$\frac{1}{z} = \frac{x - iy}{x^2 + y^2}.$$

The alternative notation z^{-1} for the reciprocal is also used.

The **quotient**, z_1/z_2, of a complex number z_1 by a non-zero complex number z_2 is

$$\frac{z_1}{z_2} = z_1\left(\frac{1}{z_2}\right).$$

Notice that $x^2 + y^2$ is strictly positive since z is non-zero.

This definition of $1/z$ works because

$$(x + iy)(x - iy) = x^2 + y^2,$$

Problem 1.2(d)

so that

$$z\left(\frac{1}{z}\right) = (x + iy)\left(\frac{x - iy}{x^2 + y^2}\right)$$

$$= \frac{(x + iy)(x - iy)}{x^2 + y^2}$$

$$= \frac{x^2 + y^2}{x^2 + y^2} = 1.$$

The above definition of quotient suggests that in order to evaluate z_1/z_2 one must first evaluate $1/z_2$ and then multiply by z_1. In practice, it is easier to do both operations at once using the following strategy.

Strategy for obtaining a quotient

To obtain the quotient

$$\frac{x_1 + iy_1}{x_2 + iy_2}, \qquad \text{where } x_2 + iy_2 \neq 0,$$

in Cartesian form, multiply both numerator and denominator by $x_2 - iy_2$, so that the denominator becomes real.

Example 1.2

Express the following numbers in Cartesian form.

(a) $\dfrac{1}{1+2i}$

(b) $\dfrac{3+4i}{1+2i}$

Solution

(a) By the strategy,

$$\frac{1}{1+2i} = \frac{1-2i}{(1+2i)(1-2i)}$$

$$= \frac{1-2i}{1+4}$$

$$= \frac{1}{5} - \frac{2}{5}i.$$

(b) By the strategy,

$$\frac{3+4i}{1+2i} = \frac{(3+4i)(1-2i)}{(1+2i)(1-2i)}$$

$$= \frac{3-2i-8i^2}{1+4}$$

$$= \frac{11-2i}{5} = \frac{11}{5} - \frac{2}{5}i. \quad \blacksquare$$

It would be acceptable to leave this solution in the form $(11-2i)/5$, since this can readily be reduced to Cartesian form.

Problem 1.3

(a) Express the following numbers in Cartesian form.

(i) $\dfrac{1}{i}$

(ii) $\dfrac{1}{1+i}$

(iii) $\dfrac{1+2i}{2+3i}$

(b) Express the quotient

$$\frac{x_1 + iy_1}{x_2 + iy_2}, \qquad \text{where } x_2 + iy_2 \neq 0,$$

in Cartesian form.

The above process of changing the sign of the imaginary part of a complex number is often used, and so we introduce the following terminology and notation.

Definition The **complex conjugate**, $\overline{z}$, of a complex number $z = x + iy$ is

$$\overline{z} = x - iy.$$

The complex conjugate of z satisfies the simple identities

$$\operatorname{Re}\overline{z} = \operatorname{Re}z \quad \text{and} \quad \operatorname{Im}\overline{z} = -\operatorname{Im}z.$$

Several more complicated identities involving complex conjugates are given in the following result.

Theorem 1.1 Properties of the complex conjugate

(a) If z is a complex number, then

    (i) $z + \overline{z} = 2\operatorname{Re} z$;

    (ii) $z - \overline{z} = 2i\operatorname{Im} z$;

    (iii) $\overline{(\overline{z})} = z$.

(b) If z_1 and z_2 are complex numbers, then

    (i) $\overline{z_1 + z_2} = \overline{z_1} + \overline{z_2}$;

    (ii) $\overline{z_1 - z_2} = \overline{z_1} - \overline{z_2}$;

    (iii) $\overline{z_1 z_2} = \overline{z_1}\,\overline{z_2}$;

    (iv) $\overline{z_1/z_2} = \overline{z_1}/\overline{z_2}$,   where $z_2 \neq 0$.

Part (b) says: 'the conjugate of a sum is the sum of the conjugates', etc.

Note the use of the long conjugate bar over expressions involving several symbols.

Proof

(a) If $z = x + iy$, then $\overline{z} = x - iy$, so that

    (i) $z + \overline{z} = (x + iy) + (x - iy) = 2x = 2\operatorname{Re} z$;

    (ii) $z - \overline{z} = (x + iy) - (x - iy) = 2iy = 2i\operatorname{Im} z$;

    (iii) $\overline{(\overline{z})} = \overline{(x - iy)} = x + iy = z$.

(b) The proofs of these identities all follow from the results of Problems 1.2 and 1.3(b). To illustrate the method we prove (iii).

Let $z_1 = x_1 + iy_1$ and $z_2 = x_2 + iy_2$, so that

$$z_1 z_2 = (x_1 x_2 - y_1 y_2) + i(x_1 y_2 + x_2 y_1), \tag{1.1}$$

by Problem 1.2(c). Also, $\overline{z_1} = x_1 - iy_1$ and $\overline{z_2} = x_2 - iy_2$, so that

$$
\begin{aligned}
\overline{z_1}\,\overline{z_2} &= (x_1 x_2 - (-y_1)(-y_2)) + i(x_1(-y_2) + x_2(-y_1)) \\
&= (x_1 x_2 - y_1 y_2) - i(x_1 y_2 + x_2 y_1) \\
&= \overline{z_1 z_2},
\end{aligned}
$$

as required. ∎

Replace y_1, y_2 by $-y_1$, $-y_2$ in Equation (1.1).

Problem 1.4

Prove the identities stated in Theorem 1.1(b), parts (i) and (iv).

Now that we have explained how to perform the usual arithmetic operations with complex numbers, it is natural to ask the following question. Do these operations have the usual properties which are known to hold for real numbers? It is a straightforward matter to check that, for example, addition of complex numbers is associative; that is, for all z_1, z_2, z_3 in $\mathbb{C}$,

$$(z_1 + z_2) + z_3 = z_1 + (z_2 + z_3).$$

It is also straightforward, but more tedious, to show that multiplication of complex numbers is associative; that is, for all z_1, z_2, z_3 in $\mathbb{C}$,

$$(z_1 z_2) z_3 = z_1 (z_2 z_3).$$

In fact, it turns out that all the usual arithmetic properties do hold for complex numbers. These are summarized in the following table.

Arithmetic in $\mathbb{C}$

Addition	Multiplication	
A1 For all z_1, z_2 in $\mathbb{C}$, $z_1 + z_2 \in \mathbb{C}$.	M1 For all z_1, z_2 in $\mathbb{C}$, $z_1 z_2 \in \mathbb{C}$.	Closure
A2 For all z in $\mathbb{C}$, $z + 0 = 0 + z = z$.	M2 For all z in $\mathbb{C}$, $z1 = 1z = z$.	Identity
A3 For all z in $\mathbb{C}$, $z + (-z) = (-z) + z = 0$.	M3 For all non-zero z in $\mathbb{C}$, $zz^{-1} = z^{-1}z = 1$.	Inverse
A4 For all z_1, z_2, z_3 in $\mathbb{C}$, $(z_1 + z_2) + z_3 = z_1 + (z_2 + z_3)$.	M4 For all z_1, z_2, z_3 in $\mathbb{C}$, $(z_1 z_2)z_3 = z_1(z_2 z_3)$.	Associative
A5 For all z_1, z_2 in $\mathbb{C}$, $z_1 + z_2 = z_2 + z_1$.	M5 For all z_1, z_2 in $\mathbb{C}$, $z_1 z_2 = z_2 z_1$.	Commutative

D   For all z_1, z_2, z_3 in $\mathbb{C}$,
$z_1(z_2 + z_3) = z_1 z_2 + z_1 z_3$.

Distributive

Once all these properties have been proved (and we shall not give the details), then the contents of the table can be described in algebraic terms as follows:

$\mathbb{C}$ is an Abelian group under the operation of *addition*, with identity 0;   A1–A5

the set of non-zero complex numbers is an Abelian group under the operation   M1–M5
of *multiplication*, with identity 1;

these two structures are linked by the *distributive* property.   D

Because $\mathbb{C}$ has all these properties, it is called a **field**; $\mathbb{Q}$ and $\mathbb{R}$ are also fields.

Notice that in property M3 we have used z^{-1} to denote the reciprocal $1/z$. It is also standard practice to use the notation z^n, where $n \in \mathbb{Z}$, for integral powers of a non-zero z; in particular, $z^0 = 1$ for all non-zero z. The zero complex number has powers $0^k = 0$ for $k = 1, 2, 3, \ldots$. We shall discuss the meaning of fractional powers, such as $z^{1/2} = \sqrt{z}$, in Section 3, and also in *Unit A2*.

For example,
$$i^2 = -1, \quad i^3 = -i, \quad i^4 = 1;$$
$$i^{-1} = -i, \quad i^{-2} = -1, \quad i^{-3} = i,$$
$$i^{-4} = 1; \quad i^0 = 1.$$

1.3 Identities with complex numbers

Because complex numbers satisfy the usual arithmetic properties, we can prove and then use all the usual algebraic identities. For example, if z_1 and z_2 are any complex numbers, then

$$(z_1 + z_2)^2 = z_1^2 + 2z_1 z_2 + z_2^2$$

and

$$z_1^2 - z_2^2 = (z_1 - z_2)(z_1 + z_2).$$

Thus, for example, if $z^2 + 9 = 0$, then

$$z^2 + 9 = (z - 3i)(z + 3i) = 0,$$

so that $z = 3i$ or $z = -3i$.

Problem 1.5

Prove the following identities.

(a) $(z_1 + z_2)^3 = z_1^3 + 3z_1^2 z_2 + 3z_1 z_2^2 + z_2^3$

(b) $z_1^3 - z_2^3 = (z_1 - z_2)(z_1^2 + z_1 z_2 + z_2^2)$

(c) $z_1^3 + z_2^3 = (z_1 + z_2)(z_1^2 - z_1 z_2 + z_2^2)$

The identities in Problem 1.5 are, in fact, special cases of two important general identities which will often be used in the course. The first of these is the Binomial Theorem, which we state in two forms. The proof is the same as in the real case, so we omit it.

Theorem 1.2 Binomial Theorem

(a) If $z \in \mathbb{C}$ and $n \in \mathbb{N}$, then

$$(1 + z)^n = \sum_{k=0}^{n} \binom{n}{k} z^k$$

$$= 1 + nz + \frac{n(n-1)}{2!} z^2 + \cdots + z^n.$$

(b) If $z_1, z_2 \in \mathbb{C}$ and $n \in \mathbb{N}$, then

$$(z_1 + z_2)^n = \sum_{k=0}^{n} \binom{n}{k} z_1^{n-k} z_2^k$$

$$= z_1^n + nz_1^{n-1} z_2 + \frac{n(n-1)}{2!} z_1^{n-2} z_2^2 + \cdots + z_2^n.$$

The binomial coefficient

$$\binom{n}{k} = \frac{n!}{k!\,(n-k)!}$$

$$= \frac{n(n-1)\ldots(n-k+1)}{k!}$$

is sometimes written as ^nC_k. Note that, by convention, $0! = 1$ and $0^0 = 1$ in these formulas.

Problem 1.5(a) is the special case with $n = 3$.

Remark It is worth recalling that the coefficients which appear in the Binomial Theorem can be arranged in the form of Pascal's Triangle, as follows.

$$
\begin{array}{cccccccccc}
(1+z)^0 & & & & & 1 & & & & \\
(1+z)^1 & & & & 1 & & 1 & & & \\
(1+z)^2 & & & 1 & & 2 & & 1 & & \\
(1+z)^3 & & 1 & & 3 & & 3 & & 1 & \\
(1+z)^4 & 1 & & 4 & & 6 & & 4 & & 1 \\
\vdots & & & & & \vdots & & & &
\end{array}
$$

Problem 1.6

Use the Binomial Theorem to simplify the following expressions.

(a) $(1 + i)^4$ (b) $(3 + 2i)^3$

Next we state the identity which is used to sum a finite geometric series. Once again we give two forms. The proof is the same as in the real case, so again we omit it.

Theorem 1.3 Geometric Series Identity

(a) If $z \in \mathbb{C}$ and $n \in \mathbb{N}$, then

$$1 - z^n = (1 - z)(1 + z + z^2 + \cdots + z^{n-1}).$$

(b) If $z_1, z_2 \in \mathbb{C}$ and $n \in \mathbb{N}$, then

$$z_1^n - z_2^n = (z_1 - z_2)(z_1^{n-1} + z_1^{n-2} z_2 + z_1^{n-3} z_2^2 + \cdots + z_2^{n-1}).$$

Problem 1.5(b) is the special case with $n = 3$.

Remark The first of these two identities can be written as

$$1 + z + z^2 + \cdots + z^n = \frac{1 - z^{n+1}}{1 - z}, \qquad \text{for } z \neq 1.$$

This is the familiar formula for summing a finite geometric series.

Problem 1.7

(a) Use the Geometric Series Identity to simplify the expression

$$1 + (1 + i) + (1 + i)^2 + (1 + i)^3.$$

(b) Use the Geometric Series Identity to find one linear factor of

$$z^5 - i.$$

    (*Hint*: $i^5 = i$.)

2 THE COMPLEX PLANE

After working through this section, you should be able to:

(a) determine the *modulus* of a given complex number;

(b) determine the *principal argument* and other *arguments* of a given non-zero complex number;

(c) convert a complex number in Cartesian form to *polar form*, and vice versa;

(d) interpret geometrically the sum, product and quotient of two complex numbers;

(e) state de Moivre's Theorem, and use it to evaluate powers of complex numbers.

2.1 Cartesian coordinates

In this section we describe a geometric interpretation of complex numbers, and we see how this interpretation leads to useful insights concerning the properties of complex numbers.

Cartesian coordinates can be used to represent the complex number $z = x + iy$ by the ordered pair (x, y) in $\mathbb{R}^2$. For example, the number $4 + 3i$ is represented by $(4, 3)$ and in Figure 2.1 this point is labelled $4 + 3i$.

Hamilton's definition of complex numbers as ordered pairs (see the Introduction) is based on this Cartesian representation.

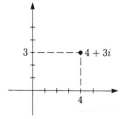

Figure 2.1

Thus we often speak of 'the point $z = x + iy$' and, with this interpretation, refer to the plane as the *complex plane* or the *z-plane*. The horizontal axis is called the *real axis* and the vertical axis is called the *imaginary axis*, and they are sometimes labelled x and y, respectively, in the usual way.

The complex plane is often called the *Argand diagram*, after Jean-Robert Argand (1768–1822), a French-Swiss mathematician, although both Gauss and Wessel (1745–1818), a Norwegian surveyor and cartographer, used the idea before Argand.

The various operations on complex numbers described in Section 1 can all be given a geometric interpretation in the complex plane. For example, if z is a complex number then, as shown in Figures 2.2 and 2.3,

$-z$ is obtained by rotating z through the angle π about the origin;

$\bar{z}$ is obtained by reflecting z in the real axis.

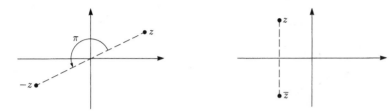

Figure 2.2   *Figure 2.3*

Since a complex number can be thought of as a based vector, the sum of two complex numbers, and also their difference, satisfy the *parallelogram law* for vectors, as shown in Figure 2.4.

The complex number $x + iy$ corresponds to the vector from the point $(0,0)$ to the point (x, y).

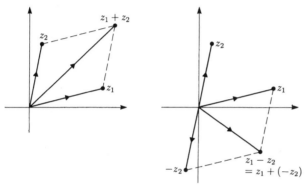

Figure 2.4

Problem 2.1

With $z_1 = 3 + i$ and $z_2 = -1 + 2i$, plot the following numbers.

(a) z_1, z_2, $-z_1$, $-z_2$, $z_1 + z_2$, $z_1 - z_2$.

(b) z_1, z_2, $\overline{z_1}$, $\overline{z_2}$, $z_1 + z_2$, $\overline{z_1 + z_2}$ (on a separate diagram).

Multiplication and division of complex numbers also have useful geometric interpretations. Before describing these, however, we need to introduce some other geometric concepts.

2.2 Polar form

The *modulus*, or *absolute value*, of a real number x is defined as follows:

$$|x| = \begin{cases} x, & x \geq 0, \\ -x, & x < 0. \end{cases}$$

Equivalently, $|x|$ is the distance along the real line from 0 to x. The modulus of a complex number z is similarly defined to be the distance from 0 to z.

Definition   The **modulus**, or **absolute value**, of a complex number $z = x + iy$ is the distance from 0 to z; it is denoted by $|z|$. Thus

$$|z| = |x + iy| = \sqrt{x^2 + y^2}.$$

For $a \geq 0$, $\sqrt{a}$ means the *non-negative* square root of a.

For example:

$$|3 + 4i| = \sqrt{3^2 + 4^2} = 5, \quad |-3| = \sqrt{(-3)^2} = 3, \quad |-2i| = \sqrt{(-2)^2} = 2.$$

These moduli are shown as distances in Figure 2.5.

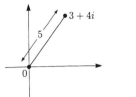
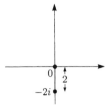

Figure 2.5

Problem 2.2

(a) Evaluate the following moduli.

    (i) $|1 + i|$ (ii) $|2 - 4i|$ (iii) $|i|$ (iv) $|-5 + 12i|$

(b) Prove that $|\overline{z}| = |z|$ and $|-z| = |z|$.

If z_1, z_2 are any two complex numbers, then, by definition, $|z_1 - z_2|$ is the distance from 0 to $z_1 - z_2$. Using the parallelogram law to add z_2 and $z_1 - z_2$ (see Figure 2.6), we deduce that

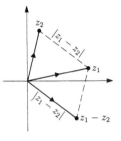

Figure 2.6

> $|z_1 - z_2|$ is the distance from z_1 to z_2.

Because $z_1 + z_2 = z_1 - (-z_2)$, there is a similar geometric interpretation for $|z_1 + z_2|$:

> $|z_1 + z_2|$ is the distance from z_1 to $-z_2$.

Problem 2.3

With $z_1 = 3 + i$ and $z_2 = -1 + 2i$, determine:

(a) $|z_1 - z_2|$;

(b) $|z_1 + z_2|$;

(c) the distance from z_2 to $-z_1$.

We now collect together various basic properties of the modulus.

Theorem 2.1 Properties of the modulus

(a) $|z| \geq 0$, with equality if and only if $z = 0$.

(b) $|\overline{z}| = |z|$ and $|-z| = |z|$.

(c) $|z|^2 = z\overline{z}$.

(d) $|z_1 - z_2| = |z_2 - z_1|$.

(e) $|z_1 z_2| = |z_1||z_2|$ and $|z_1/z_2| = |z_1|/|z_2|$, where $z_2 \neq 0$.

Property (d) says, algebraically, that the distance from z_1 to z_2 is the same as the distance from z_2 to z_1.

16

Proof Property (a) follows from the fact that $|z| = \sqrt{x^2 + y^2}$, if $z = x + iy$, and property (b) was proved in Problem 2.2(b). To prove property (c), note that if $z = x + iy$ then

$$z\bar{z} = (x + iy)(x - iy) = x^2 + y^2 = |z|^2.$$

Property (d) follows from property (b), since $z_2 - z_1 = -(z_1 - z_2)$.

Each of the identities in property (e) can be proved by writing $z_1 = x_1 + iy_1$, $z_2 = x_2 + iy_2$ and then calculating both sides. However, it is neater to use property (c) and Theorem 1.1, as follows:

$$
\begin{aligned}
|z_1 z_2|^2 &= (z_1 z_2)\overline{(z_1 z_2)} && \text{(property (c))} \\
&= (z_1 z_2)(\overline{z_1}\,\overline{z_2}) && \text{(Theorem 1.1(b), part (iii))} \\
&= (z_1 \overline{z_1})(z_2 \overline{z_2}) && \text{(associativity and commutativity)} \\
&= |z_1|^2 |z_2|^2 && \text{(property (c)),}
\end{aligned}
$$

so that $|z_1 z_2| = |z_1||z_2|$. Similarly, if $z_2 \neq 0$ (so that $\overline{z_2} \neq 0$ and $|z_2| > 0$), then

$$|z_1/z_2|^2 = (z_1/z_2)\overline{(z_1/z_2)} = (z_1/z_2)(\overline{z_1}/\overline{z_2}) = (z_1\overline{z_1})/(z_2\overline{z_2}) = |z_1|^2/|z_2|^2,$$

so that $|z_1/z_2| = |z_1|/|z_2|$. ■

If the modulus $|z|$ of a complex number z is equal to 0, then z itself must equal 0 (and vice versa). However, the modulus of a *non-zero* complex number does not determine the number completely; all the points which lie on the circle of radius r centred at the origin have the same modulus, namely r. We can determine the non-zero complex number z completely by giving its modulus $|z| = r$ together with:

> the angle θ that the line from the origin to z makes with the positive real axis.

Angles used to determine position in this way are conventionally taken to be positive when measured in an anticlockwise direction from the positive real axis, and negative when measured in a clockwise direction.

For example, $1 + i$ has modulus $\sqrt{2}$ and the (positive) angle that the line from the origin to $1 + i$ makes with the positive real axis is $\pi/4$ (see Figure 2.7). Of course, $\pi/4$ is not the only angle which, along with the modulus $\sqrt{2}$, specifies $1 + i$; any one of the angles

$$\ldots, \quad \frac{\pi}{4} - 2\pi, \quad \frac{\pi}{4}, \quad \frac{\pi}{4} + 2\pi, \quad \frac{\pi}{4} + 4\pi, \quad \ldots$$

would do just as well — in particular the (negative) angle $\pi/4 - 2\pi = -7\pi/4$ (see Figure 2.8). This feature is reflected in the following definition by the use of the sine and cosine functions. Figure 2.9 illustrates the definition, showing one argument of $z = x + iy$.

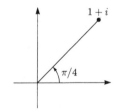

Figure 2.7

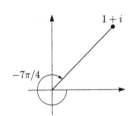

Figure 2.8

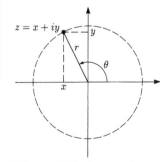

Figure 2.9 An argument of z

Definition An **argument** of a non-zero complex number $z = x + iy$ with $|z| = r$ is an angle θ (measured in radians) such that

$$\cos\theta = \frac{x}{r} \quad \text{and} \quad \sin\theta = \frac{y}{r}.$$

Remarks

1 No argument is assigned to 0.

2 Each non-zero complex number has infinitely many arguments, all differing by integer multiples of 2π. For example, the arguments of $1 + i$ (see above) are

$$\ldots, \quad -7\pi/4, \quad \pi/4, \quad 9\pi/4, \quad 17\pi/4, \quad \ldots,$$

which may be written as

$$\pi/4 + 2k\pi, \quad \text{where } k \in \mathbb{Z}.$$

3 For some complex numbers, arguments are easily obtained by plotting the point. For example, Figure 2.10 shows that $3\pi/4$ is an argument of $-1+i$, $\pi/2$ is an argument of i and $-\pi/4$ is an argument of $1-i$. The *calculation* of arguments is dealt with later in the section.

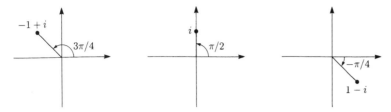

Figure 2.10

Since any non-zero complex number is completely determined by its modulus and any one of its arguments, these two quantities can be used to define an alternative coordinate system for non-zero complex numbers.

> **Definition**   The ordered pair (r, θ), where r is the modulus of a non-zero complex number z and θ is an argument of z, are called **polar coordinates** of z. The expression
>
> $$z = r(\cos\theta + i\sin\theta)$$
>
> is called a representation of z in **polar form**.

Note that $r > 0$ and $\theta \in \mathbb{R}$.

Remarks

1 Some alternative notations for polar coordinates are $\langle r, \theta \rangle$ and $[r, \theta]$. We shall rarely use polar coordinates, preferring almost always to use polar form.

2 It follows from the definition of polar form that if $z = x + iy$, then

$$x = r\cos\theta \quad \text{and} \quad y = r\sin\theta.$$

Example 2.1

Represent $-1-i$ in polar form.

Solution

Here

$$r = |-1-i| = \sqrt{(-1)^2 + (-1)^2} = \sqrt{2}$$

and, from Figure 2.11, one choice for θ is $5\pi/4$. Thus

$$-1-i = \sqrt{2}(\cos 5\pi/4 + i\sin 5\pi/4)$$

is in polar form.   ■

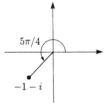

Figure 2.11

Another polar form for $-1-i$ is
$$\sqrt{2}(\cos(-3\pi/4) + i\sin(-3\pi/4)).$$

Problem 2.4 _____

(a) Represent the complex number i in polar form.

(b) Represent each of the following complex numbers in Cartesian form.

    (i) $2(\cos\pi/3 + i\sin\pi/3)$

    (ii) $3(\cos(-\pi/4) + i\sin(-\pi/4))$

The terminology 'arg z' is often used to denote an argument of a non-zero complex number z. Without further information, however, the expression arg z is ambiguous, since z has infinitely many arguments, and so we shall use it rarely. Instead, we select one argument for special attention and call this the *principal argument* (a shortened version of the more conventional 'principal value of the argument').

> **Definition** The **principal argument** of a non-zero complex number z is the unique argument θ of z satisfying $-\pi < \theta \le \pi$; it is denoted by
>
> $\qquad \theta = \operatorname{Arg} z.$
>
> (Note the capital A in Arg.)

Since the arguments of a non-zero complex number differ by multiples of 2π, exactly *one* of them satisfies $-\pi < \theta \le \pi$.

For example, as you have seen, the arguments of $1 + i$ are

$\qquad \ldots, \quad -7\pi/4, \quad \pi/4, \quad 9\pi/4, \quad 17\pi/4, \quad \ldots,$

and hence $\operatorname{Arg}(1 + i) = \pi/4$ because $-\pi < \pi/4 \le \pi$.

For complex numbers z such as $1 + i$ it is easy to determine $\operatorname{Arg} z$ by inspection. In general, the following strategy may be applied.

There are other equally valid strategies.

Strategy for determining principal arguments

To determine the principal argument θ of a non-zero complex number $z = x + iy$.

Case (i) If z lies on one of the axes, then θ is evident (see Figure 2.12).

Case (ii) If z does not lie on one of the axes, then

(a) decide in which quadrant z lies (by plotting z if necessary), and then calculate the angle

$\qquad \phi = \tan^{-1}(|y|/|x|)$

(see Figure 2.13);

(b) obtain θ in terms of ϕ by using the appropriate formula in Figure 2.14.

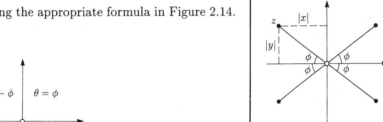

Figure 2.12

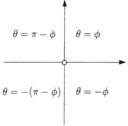

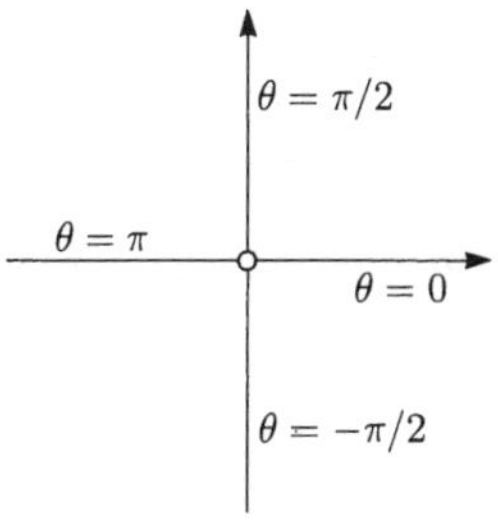

Figure 2.13

$\theta = \pi - \phi \qquad \theta = \phi$

$\theta = -(\pi - \phi) \qquad \theta = -\phi$

Figure 2.14

Remarks

1 Having found $\operatorname{Arg} z$, other arguments may be obtained by adding integer multiples of 2π to it.

2 The requirement that the principal argument $\operatorname{Arg} z$ should satisfy $-\pi < \operatorname{Arg} z \le \pi$ may seem rather arbitrary. It is, but if some choice has to be made, then this one is marginally better than others. One reason for the choice $-\pi < \operatorname{Arg} z \le \pi$ will become clear in *Units A3* and *A4*.

Some texts prefer $0 \le \operatorname{Arg} z < 2\pi$.

Example 2.2

Find the principal argument of each of the following complex numbers.

(a) $1 + 2i$ (b) $-1 - \sqrt{3}i$ (c) $-1 + \sqrt{3}i$

Solution

Following the strategy (case (ii) each time), we have the following.

(a) $1 + 2i$ lies in the first quadrant (Figure 2.15), and

$$\phi = \tan^{-1}(2/1) = \tan^{-1} 2;$$

thus the principal argument θ is

$$\begin{aligned} \theta &= \phi \qquad \text{(Figure 2.14)} \\ &= \tan^{-1} 2 \qquad \text{(about 1.11 radians).} \end{aligned}$$

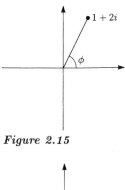

Figure 2.15

(b) $-1 - \sqrt{3}i$ lies in the third quadrant (Figure 2.16), and

$$\phi = \tan^{-1}(|-\sqrt{3}|/|-1|) = \tan^{-1}\sqrt{3} = \pi/3;$$

thus the principal argument θ is

$$\begin{aligned} \theta &= -(\pi - \pi/3) \qquad \text{(Figure 2.14)} \\ &= -2\pi/3. \end{aligned}$$

Figure 2.16

(c) $-1 + \sqrt{3}i$ lies in the second quadrant (Figure 2.17), and

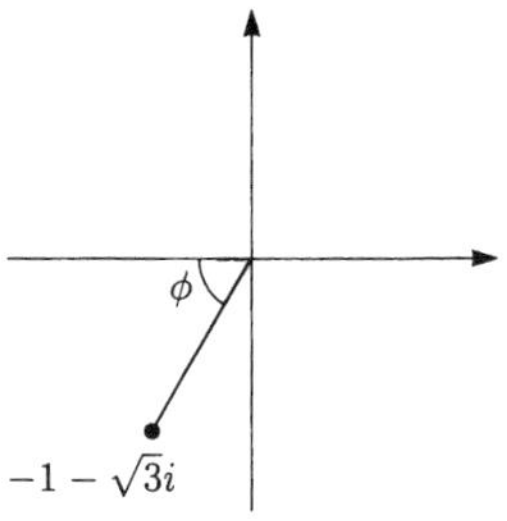

$$\phi = \tan^{-1}(\sqrt{3}/|-1|) = \tan^{-1}\sqrt{3} = \pi/3;$$

thus the principal argument θ is

$$\begin{aligned} \theta &= \pi - \phi \qquad \text{(Figure 2.14)} \\ &= 2\pi/3. \end{aligned}$$

(Of course, knowing that $\text{Arg}(-1 - \sqrt{3}i) = -2\pi/3$ from part (b) and that $-1 + \sqrt{3}i$ is the conjugate of $-1 - \sqrt{3}i$, it follows immediately that

$$\text{Arg}(-1 + \sqrt{3}i) = -(-2\pi/3) = 2\pi/3). \quad ■$$

Figure 2.17

Problem 2.5

For each of the following complex numbers z, write down $\text{Arg}\, z$ and express z in polar form.

(a) -4 (b) $3\sqrt{3} + 3i$ (c) $\sqrt{3} - i$ (d) $-1 - i$

2.3 The geometric interpretation of multiplication and division

A geometric interpretation of the multiplication of complex numbers can be given using the polar form of complex numbers. Indeed, if z_1 and z_2 are non-zero complex numbers with polar forms

$$z_1 = r_1(\cos\theta_1 + i\sin\theta_1) \quad \text{and} \quad z_2 = r_2(\cos\theta_2 + i\sin\theta_2),$$

then

$$\begin{aligned} z_1 z_2 &= r_1 r_2(\cos\theta_1 + i\sin\theta_1)(\cos\theta_2 + i\sin\theta_2) \\ &= r_1 r_2((\cos\theta_1\cos\theta_2 - \sin\theta_1\sin\theta_2) + i(\sin\theta_1\cos\theta_2 + \cos\theta_1\sin\theta_2)) \\ &= r_1 r_2(\cos(\theta_1 + \theta_2) + i\sin(\theta_1 + \theta_2)), \end{aligned}$$

by using the formulas for the sine and cosine of the sum of two angles. The above formula shows that $|z_1 z_2| = r_1 r_2 = |z_1||z_2|$, which we knew already, and also that the number $\theta_1 + \theta_2$ is an argument of $z_1 z_2$. Thus we can describe the effect of multiplying z_1 by z_2 (both non-zero) as follows.

> The modulus of $z_1 z_2$ is the modulus of z_1 *multiplied* by the modulus of z_2; an argument of $z_1 z_2$ is an argument of z_1 *plus* an argument of z_2.

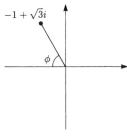

Thus the geometric effect on z_1 of multiplying it by z_2 is to scale it by the factor $|z_2|$ and rotate it about 0 through the angle $\operatorname{Arg} z_2$. (This rotation is anticlockwise if $\operatorname{Arg} z_2 > 0$, clockwise if $\operatorname{Arg} z_2 < 0$.) This is illustrated in Figure 2.18 for the case where z_1, z_2 and $z_1 z_2$ are in the first quadrant, θ_1, θ_2 and $\theta_1 + \theta_2$ being their principal arguments. (Note that in Figure 2.18 we have omitted the arrowheads from the arguments. Henceforth, this will be our usual practice.)

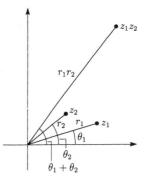

Figure 2.18

Unfortunately, it is not always true that the principal argument of $z_1 z_2$ is the sum of the principal arguments of z_1 and z_2. It may differ from this sum by $\pm 2\pi$. For example, if $\operatorname{Arg} z_1 = \pi/2$ and $\operatorname{Arg} z_2 = 3\pi/4$, then

$$\operatorname{Arg} z_1 + \operatorname{Arg} z_2 = 5\pi/4.$$

Thus $5\pi/4$ is an argument of $z_1 z_2$ but (because $5\pi/4 > \pi$) it is not the principal argument of $z_1 z_2$. In fact, since $-\pi < \operatorname{Arg}(z_1 z_2) \leq \pi$,

$$\operatorname{Arg}(z_1 z_2) = 5\pi/4 - 2\pi = -3\pi/4.$$

On the other hand, if $\operatorname{Arg} z_1 = -\pi/4$ and $\operatorname{Arg} z_2 = -7\pi/8$, then

$$\operatorname{Arg} z_1 + \operatorname{Arg} z_2 = -9\pi/8,$$

which is *an* argument of $z_1 z_2$, but

$$\operatorname{Arg}(z_1 z_2) = -9\pi/8 + 2\pi = 7\pi/8.$$

In general, since $-2\pi < \operatorname{Arg} z_1 + \operatorname{Arg} z_2 \leq 2\pi$, we have the following property of $\operatorname{Arg} z$.

If z_1 and z_2 are (non-zero) complex numbers, then

$$\operatorname{Arg}(z_1 z_2) = \operatorname{Arg} z_1 + \operatorname{Arg} z_2 + 2n\pi,$$

where n is -1, 0 or 1 according as $\operatorname{Arg} z_1 + \operatorname{Arg} z_2$ is greater than π, lies in the interval $]-\pi, \pi]$, or is less than or equal to $-\pi$.

Problem 2.6

Use polar forms of the complex numbers

$$z_1 = -1 - \sqrt{3}i, \quad z_2 = 3\sqrt{3} + 3i,$$

to evaluate $z_1 z_2$ and z_1^2. (You will find Example 2.2(b) and Problem 2.5(b) useful.)

Problem 2.7

Describe the geometric effect on a complex number z of multiplying z by $2i$.

As you might expect, the polar form of complex numbers is useful also for division. Indeed, if z_1 and z_2 are non-zero complex numbers with polar forms

$$z_1 = r_1(\cos\theta_1 + i\sin\theta_1) \quad \text{and} \quad z_2 = r_2(\cos\theta_2 + i\sin\theta_2),$$

then

$$
\begin{aligned}
\frac{z_1}{z_2} &= \frac{r_1(\cos\theta_1 + i\sin\theta_1)}{r_2(\cos\theta_2 + i\sin\theta_2)} \\
&= \frac{r_1}{r_2}\frac{(\cos\theta_1 + i\sin\theta_1)(\cos\theta_2 - i\sin\theta_2)}{(\cos\theta_2 + i\sin\theta_2)(\cos\theta_2 - i\sin\theta_2)} \\
&= \frac{r_1}{r_2}\frac{((\cos\theta_1\cos\theta_2 + \sin\theta_1\sin\theta_2) + i(\sin\theta_1\cos\theta_2 - \cos\theta_1\sin\theta_2))}{\cos^2\theta_2 + \sin^2\theta_2} \\
&= \frac{r_1}{r_2}(\cos(\theta_1 - \theta_2) + i\sin(\theta_1 - \theta_2)), \qquad\qquad (2.1)
\end{aligned}
$$

using the formulas for the sine and cosine of the difference of two angles. This formula shows that $|z_1/z_2| = r_1/r_2 = |z_1|/|z_2|$ and also that the number $\theta_1 - \theta_2$

is an argument of z_1/z_2. Thus we can describe the effect of dividing non-zero complex numbers as follows.

The modulus of z_1/z_2 is the modulus of z_1 *divided* by the modulus of z_2; an argument of z_1/z_2 is an argument of z_1 *minus* an argument of z_2.

Thus the geometric effect on z_1 of dividing it by z_2 is to scale it by the factor $1/|z_2|$ and rotate it about 0 through the angle $-\operatorname{Arg} z_2$. (This rotation is clockwise if $\operatorname{Arg} z_2 > 0$, anticlockwise if $\operatorname{Arg} z_2 < 0$.)

Problem 2.8

Use polar forms of the complex numbers

$$z_1 = 1 + \sqrt{3}i, \quad z_2 = \sqrt{3} - i,$$

to evaluate z_1/z_2.

Problem 2.9

Describe the geometric effect on a complex number z of dividing z by $2i$.

An important special case of Formula (2.1) for the quotient z_1/z_2 is obtained when

$$z_1 = 1 \quad \text{and} \quad z_2 = r(\cos\theta + i\sin\theta),$$

so that

$$r_1 = 1, \quad \theta_1 = 0, \quad r_2 = r, \quad \theta_2 = \theta.$$

In this case we find that

$$\frac{1}{r(\cos\theta + i\sin\theta)} = \frac{1(\cos 0 + i\sin 0)}{r(\cos\theta + i\sin\theta)}$$

$$= \frac{1}{r}(\cos(0 - \theta) + i\sin(0 - \theta))$$

$$= \frac{1}{r}(\cos(-\theta) + i\sin(-\theta)).$$

Thus the reciprocal of a non-zero complex number z can be described as follows.

The modulus of z^{-1} is the *reciprocal* of the modulus of z; an argument of z^{-1} is the *negative* of an argument of z.

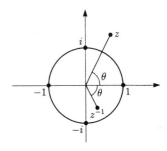

Figure 2.19

Notice that if z lies outside the circle of radius 1 centred at 0, so that $|z| > 1$, then z^{-1} lies inside this circle (because $|z^{-1}| < 1$), as shown in Figure 2.19, and vice versa. If z lies on this circle, then z is of the form $z = \cos\theta + i\sin\theta$ and

$$z^{-1} = (\cos\theta + i\sin\theta)^{-1}$$

$$= \cos(-\theta) + i\sin(-\theta),$$

so z^{-1} also lies on the circle (see Figure 2.20); moreover, in this case

$$z^{-1} = \cos(-\theta) + i\sin(-\theta)$$

$$= \cos\theta - i\sin\theta$$

$$= \overline{z}.$$

In general, for all non-zero z,

$$z^{-1} = \frac{\overline{z}}{z\overline{z}} = \frac{1}{|z|^2}\overline{z},$$

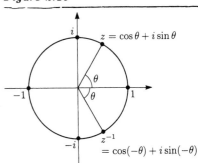

Figure 2.20

and so (since $1/|z|^2$ is real)
$$\mathrm{Arg}\, z^{-1} = \mathrm{Arg}\, \overline{z}.$$
Also, if $-\pi < \mathrm{Arg}\, z < \pi$, then
$$\mathrm{Arg}\, \overline{z} = -\,\mathrm{Arg}\, z,$$
since $\overline{z}$ is the reflection of z in the real axis. Thus we have the following properties of $\mathrm{Arg}\, z$.

If z is non-zero and $-\pi < \mathrm{Arg}\, z < \pi$, then
$$\mathrm{Arg}\, \overline{z} = \mathrm{Arg}\, z^{-1} = -\,\mathrm{Arg}\, z.$$

Problem 2.10

Use a polar form of $1 + i$ to evaluate $(1 + i)^{-1}$.

The product of several complex numbers $z_1, z_2, \ldots, z_n$ has a similar interpretation.

The modulus of $z_1 z_2 \ldots z_n$ is the *product* of the moduli of $z_1, z_2, \ldots, z_n$; an argument of $z_1 z_2 \ldots z_n$ is the *sum* of arguments of $z_1, z_2, \ldots, z_n$.

In other words, the product of the n complex numbers
$$z_k = r_k(\cos\theta_k + i\sin\theta_k), \qquad k = 1, 2, \ldots, n,$$
is given by
$$\begin{aligned}
z_1 z_2 \ldots z_n = r_1 r_2 \ldots r_n (&\cos(\theta_1 + \theta_2 + \cdots + \theta_n) \\
&+ i\sin(\theta_1 + \theta_2 + \cdots + \theta_n)).
\end{aligned} \tag{2.2}$$

Problem 2.11

Use polar forms of the complex numbers
$$z_1 = 1 + i, \quad z_2 = 1 + \sqrt{3}i, \quad z_3 = \sqrt{3} + i,$$
to evaluate $z_1 z_2 z_3$.

In the next subsection, polar form is used to calculate powers.

2.4 de Moivre's Theorem

An important special case of Formula (2.2) for the product $z_1 z_2 \ldots z_n$ is obtained when
$$z_1 = z_2 = \ldots = z_n = \cos\theta + i\sin\theta,$$
so that
$$r_1 = r_2 = \ldots = r_n = 1 \quad \text{and} \quad \theta_1 = \theta_2 = \ldots = \theta_n = \theta.$$
In this case, Formula (2.2) becomes
$$(\cos\theta + i\sin\theta)^n = \cos n\theta + i\sin n\theta, \qquad n = 1, 2, \ldots;$$
this identity is due to de Moivre.

Abraham de Moivre, who worked mostly in England, was an 18th century probabilist. He knew this identity as early as 1707.

Figure 2.21 shows the geometric interpretation of de Moivre's identity. The powers of $\cos\theta + i\sin\theta$ are equally spaced around the circle with centre 0 and radius 1, the angle between adjacent powers being θ. Each multiplication by $\cos\theta + i\sin\theta$ gives rise to a rotation through θ (radians) about 0.

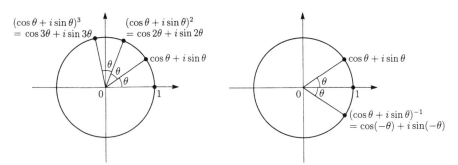

Figure 2.21   *Figure 2.22*

In Figure 2.22, the position of $(\cos\theta + i\sin\theta)^{-1}$ suggests that de Moivre's identity holds also for negative integer powers; we now show that this is true.

Theorem 2.2 de Moivre's Theorem

If n is an integer and θ is a real number, then

$$(\cos\theta + i\sin\theta)^n = \cos n\theta + i\sin n\theta.$$

Proof   We have already proved de Moivre's Theorem for a positive integer n, and it is also true for:

$$n = 0, \quad \text{since } (\cos\theta + i\sin\theta)^0 = 1 = \cos 0 + i\sin 0,$$

and

$$n = -1, \quad \text{since } (\cos\theta + i\sin\theta)^{-1} = \cos(-\theta) + i\sin(-\theta).$$

To complete the proof, note that if m is a positive integer, then

$$
\begin{aligned}
(\cos\theta + i\sin\theta)^{-m} &= ((\cos\theta + i\sin\theta)^{-1})^m \\
&= (\cos(-\theta) + i\sin(-\theta))^m \\
&= \cos(-m\theta) + i\sin(-m\theta).
\end{aligned}
$$

Hence de Moivre's Theorem holds also if $n = -m$, where m is a positive integer.   ∎

Problem 2.12 _____

Use de Moivre's Theorem to evaluate the following powers.

(a) $(\sqrt{3} + i)^4$ (b) $(1 - \sqrt{3}i)^3$ (c) $(1 + i)^{10}$

(d) $(-1 + i)^{-8}$ (e) $(\sqrt{3} + i)^{-6}$

3 SOLVING EQUATIONS WITH COMPLEX NUMBERS

After working through this section, you should be able to:

(a) calculate the *nth roots* of a complex number;

(b) solve certain polynomial equations with complex coefficients.

As we explained in the Introduction, the use of complex numbers allows both quadratic and cubic equations with real coefficients to be solved. You will see in this course that complex numbers enable us to solve many equations which may not have real solutions. In this section we describe various polynomial equations whose complex solutions can be found explicitly.

3.1 Calculating *n*th roots

If a is a non-negative real number and n is a positive integer, then $\sqrt[n]{a}$ or $a^{1/n}$ denotes the non-negative nth root of a; that is, the unique non-negative number x such that $x^n = a$. In this subsection, we discuss the nth roots of a complex number, beginning with square roots.

The simplest quadratic equation which has a complex solution but no real solutions is

$$z^2 + 1 = 0, \qquad \text{that is, } z^2 = -1.$$

One solution to this equation is $z = i$, since $i^2 = -1$; another solution is $z = -i$, since $(-i)^2 = (-1)^2 i^2 = -1$.

A more general quadratic equation is

$$z^2 = w, \tag{3.1}$$

where w is a given complex number. Any solution z of Equation (3.1) is called a **square root** of w; for example, both i and $-i$ are square roots of -1.

In fact, we shall show shortly that each non-zero complex number w has exactly two square roots. The following example shows how to find square roots geometrically.

Later we shall introduce $\sqrt{w}$ or $w^{1/2}$ to denote a particular square root of w.

Example 3.1

Find (the) two solutions of the equation

$$z^2 = i.$$

Solution

By the geometric properties of complex multiplication, described in Subsection 2.3,

the modulus of z^2 is the square of the modulus of z

and

an argument of z^2 is double an argument of z.

Since i has modulus 1 and argument $\pi/2$, one solution of $z^2 = i$ is obtained by taking z to have modulus $\sqrt{1} = 1$ and argument $\frac{1}{2}(\pi/2) = \pi/4$ (see Figure 3.1). This gives

$$z = 1\left(\cos\frac{\pi}{4} + i\sin\frac{\pi}{4}\right)$$
$$= \frac{1}{\sqrt{2}} + \frac{1}{\sqrt{2}}i.$$

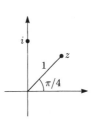

Figure 3.1

Check:

$$\left(\frac{1}{\sqrt{2}} + \frac{1}{\sqrt{2}}i\right)^2$$
$$= \frac{1}{2} + 2\frac{1}{\sqrt{2}}\frac{1}{\sqrt{2}}i - \frac{1}{2} = i.$$

25

Since $\dfrac{1}{\sqrt{2}} + \dfrac{1}{\sqrt{2}}i$ is a solution of $z^2 = i$ and $(-1)^2 = 1$,

$$z = -\left(\frac{1}{\sqrt{2}} + \frac{1}{\sqrt{2}}i\right)$$

is another solution of $z^2 = i$. Therefore the required solutions are

$$z = \pm\left(\frac{1}{\sqrt{2}} + \frac{1}{\sqrt{2}}i\right),$$

illustrated in Figure 3.2. ∎

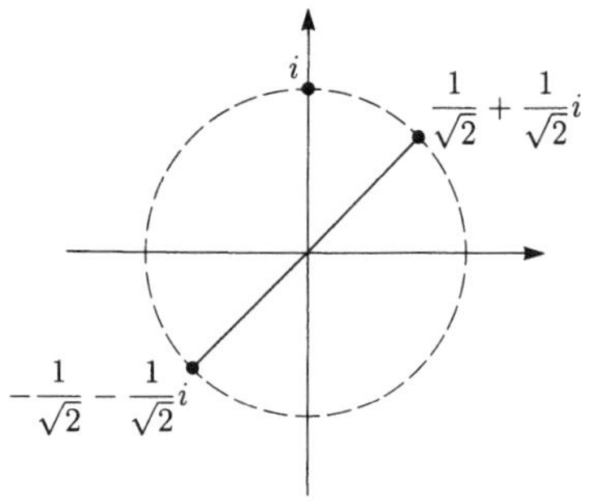

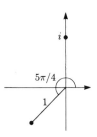

Figure 3.2

Remarks

1 Notice that the second solution could also have been found geometrically. If we had begun by taking i to have modulus 1 and argument $\pi/2 + 2\pi = 5\pi/2$, then the corresponding solution z would have modulus $\sqrt{1} = 1$ and argument $\frac{1}{2}(5\pi/2) = 5\pi/4$ (see Figure 3.3). This gives

$$z = 1\left(\cos\frac{5\pi}{4} + i\sin\frac{5\pi}{4}\right) = -\frac{1}{\sqrt{2}} - \frac{1}{\sqrt{2}}i.$$

2 Note that if w is any complex number and z is a square root of w, then $-z$ is also a square root of w.

3 An alternative method of solving $z^2 = i$ is to write $z = x + iy$, equate the real parts and imaginary parts of

$$(x + iy)^2 = x^2 - y^2 + 2xy\,i = i,$$

and then solve the resulting equations for x and y (see Exercise 3.2). This method is, however, not suitable for finding the nth roots of complex numbers if $n > 2$.

Figure 3.3

Recall that $z_1 = z_2$ means that
  $\text{Re}\,z_1 = \text{Re}\,z_2$
and
  $\text{Im}\,z_1 = \text{Im}\,z_2.$

Problem 3.1

Find (the) two solutions of the equation

$$z^2 = -1 + \sqrt{3}i.$$

We now turn to the more general equation

$$z^n = w,$$

where w is a given complex number and n is any positive integer with $n \geq 2$. Each solution of $z^n = w$ is called an **nth root** of w. We shall show shortly that each non-zero complex number w has exactly n nth roots.

As a simple example, consider the equation

$$z^3 = -8,$$

which has just one real solution, $z = -2$. To discover other solutions, we put -8 in polar form. Since

$$-8 = 8(\cos\pi + i\sin\pi),$$

a solution of $z^3 = -8$ is obtained by taking z to have modulus $\sqrt[3]{8} = 2$ and argument $\pi/3$. This gives

$$z = 2(\cos\pi/3 + i\sin\pi/3) = 1 + \sqrt{3}i.$$

Another polar form of -8 is

$$-8 = 8(\cos(-\pi) + i\sin(-\pi)),$$

and so

$$z = 2(\cos(-\pi/3) + i\sin(-\pi/3)) = 1 - \sqrt{3}i$$

also satisfies $z^3 = -8$. Thus (the) three solutions of $z^3 = -8$ are

$$z = -2,\ 1 + \sqrt{3}i,\ 1 - \sqrt{3}i.$$

If $w = 0$, then $z = 0$ is the only solution.

For $a \geq 0$, $\sqrt[3]{a}$ means the non-negative cube root of a.

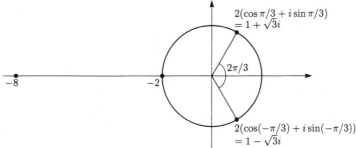

Figure 3.4

Notice that these solutions all lie on the circle with centre 0 and radius 2 (see Figure 3.4), and that the angle between adjacent solutions is $2\pi/3$. Thus these three cube roots form the vertices of an equilateral triangle. This is a special case of the following general result for nth roots.

Theorem 3.1 Let

$$w = \rho(\cos\phi + i\sin\phi)$$

be a non-zero complex number in polar form. Then w has exactly n nth roots, given by

$$z_k = \rho^{1/n}\left(\cos\left(\frac{\phi}{n} + k\frac{2\pi}{n}\right) + i\sin\left(\frac{\phi}{n} + k\frac{2\pi}{n}\right)\right),$$
$$k = 0, 1, \ldots, n-1.$$

These n nth roots form the vertices of an n-sided regular polygon inscribed in the circle of radius $\rho^{1/n}$ centred at 0.

We are using the Greek letters ρ and ϕ here because r and θ are needed in the proof.

Proof We seek the solutions of $z^n = w$ in polar form, $z = r(\cos\theta + i\sin\theta)$. Since $w = \rho(\cos\phi + i\sin\phi)$, the equation $z^n = w$ takes the form

$$r^n(\cos\theta + i\sin\theta)^n = \rho(\cos\phi + i\sin\phi);$$

that is,

$$r^n(\cos n\theta + i\sin n\theta) = \rho(\cos\phi + i\sin\phi),$$

by de Moivre's Theorem.

Equating the moduli of both sides, and using the fact that the arguments of the two sides differ by an integer multiple of 2π, we obtain

$$r^n = \rho \quad \text{and} \quad n\theta = \phi + 2k\pi, \quad \text{where } k \in \mathbb{Z}.$$

The only possible value of r is $\rho^{1/n}$ (since r must be non-negative), and the only possible values of θ are

$$\theta = \frac{\phi}{n} + k\frac{2\pi}{n}, \quad \text{where } k \in \mathbb{Z}.$$

Hence the solutions of $z^n = w$ are all of the form

$$z_k = \rho^{1/n}\left(\cos\left(\frac{\phi}{n} + k\frac{2\pi}{n}\right) + i\sin\left(\frac{\phi}{n} + k\frac{2\pi}{n}\right)\right), \quad \text{where } k \in \mathbb{Z}.$$

At first sight, it might appear that we have found infinitely many solutions, one for each value of k. However, not all these solutions are distinct. Indeed, if k_1 and k_2 differ by an integer multiple of n, say

$$k_2 = k_1 + mn, \quad \text{where } m \in \mathbb{Z},$$

then

$$\frac{\phi}{n} + k_2\frac{2\pi}{n} = \frac{\phi}{n} + (k_1 + mn)\frac{2\pi}{n} = \left(\frac{\phi}{n} + k_1\frac{2\pi}{n}\right) + 2\pi m,$$

Sometimes the loose phrase 'equating moduli and arguments' is used.

For $a \geq 0$, $a^{1/n}$ means the non-negative nth root of a.

so that $\dfrac{\phi}{n} + k_2 \dfrac{2\pi}{n}$ and $\dfrac{\phi}{n} + k_1 \dfrac{2\pi}{n}$ differ by an integer multiple of 2π. Hence the solutions arising from k_1 and k_2 are identical and so all possible solutions of $z^n = w$ arise from the integers $k = 0, 1, \dots, n-1$. These n solutions are clearly distinct, since they lie on the circle of radius $\rho^{1/n}$ centred at 0, with the angle $2\pi/n$ between adjacent solutions. Thus they do form the vertices of a regular n-sided polygon (Figure 3.5 illustrates this in the case $n = 6$). ∎

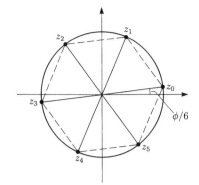

Figure 3.5

If $w = \rho(\cos\phi + i\sin\phi)$, where ϕ is the *principal argument* of w, then

$$z_0 = \rho^{1/n}\left(\cos\frac{\phi}{n} + i\sin\frac{\phi}{n}\right)$$

is called the **principal nth root** of w, denoted by $\sqrt[n]{w}$ or $w^{1/n}$. Note that if w is a positive real number, with principal argument $\phi = 0$, then the principal nth root of w has argument 0 and so is positive. Hence this use of the notation $\sqrt[n]{w}$ is consistent with the familiar real case. This consistency is taken further because for $0 \in \mathbb{C}$, $\sqrt[n]{0}$ or $0^{1/n}$ is defined to be 0.

In particular, the principal square root of w is denoted by $\sqrt{w}$ or $w^{1/2}$.

A particularly important case of Theorem 3.1 occurs when $w = 1$, so that $\rho = 1$ and $\phi = 0$.

Corollary   The number 1 has exactly n nth roots, given by

$$z_k = \cos\left(\frac{2\pi k}{n}\right) + i\sin\left(\frac{2\pi k}{n}\right), \qquad k = 0, 1, \dots, n-1.$$

These are called the **nth roots of unity**.

Note that $z_0 = 1$ is the principal nth root of unity for each n.

The nth roots of unity lie on the circle of radius 1 centred at 0, with the angle $2\pi/n$ between adjacent roots. The cases $n = 2, 3, 4$ are illustrated in Figure 3.6.

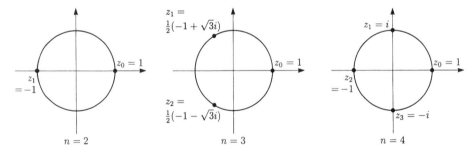

Figure 3.6   Roots of unity

Example 3.2

Determine the fourth roots of $-8 + 8\sqrt{3}i$ in polar and Cartesian forms, plot them in the complex plane, and indicate the principal fourth root.

Solution

Since $|-8 + 8\sqrt{3}i| = \sqrt{(-8)^2 + (8\sqrt{3})^2} = 16$ and $-8 + 8\sqrt{3}i$ has principal

argument $\pi - \tan^{-1}\dfrac{8\sqrt{3}}{8} = \pi - \dfrac{\pi}{3} = \dfrac{2\pi}{3}$, we deduce that

$$-8 + 8\sqrt{3}i = 16\left(\cos\frac{2\pi}{3} + i\sin\frac{2\pi}{3}\right).$$

Hence, by Theorem 3.1 with $\rho = 16$ and $\phi = 2\pi/3$, the four fourth roots of $-8 + 8\sqrt{3}i$ are

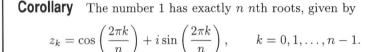
$$z_k = 16^{1/4}\left(\cos\left(\frac{2\pi/3}{4} + k\frac{2\pi}{4}\right) + i\sin\left(\frac{2\pi/3}{4} + k\frac{2\pi}{4}\right)\right)$$
$$= 2\left(\cos\left(\frac{\pi}{6} + k\frac{\pi}{2}\right) + i\sin\left(\frac{\pi}{6} + k\frac{\pi}{2}\right)\right), \qquad k = 0, 1, 2, 3.$$

Thus the arguments of the four fourth roots of $-8 + 8\sqrt{3}i$ are

$$\frac{\pi}{6}, \quad \frac{\pi}{6} + \frac{\pi}{2} = \frac{2\pi}{3}, \quad \frac{\pi}{6} + 2\left(\frac{\pi}{2}\right) = \frac{7\pi}{6}, \quad \frac{\pi}{6} + 3\left(\frac{\pi}{2}\right) = \frac{5\pi}{3},$$

and so the polar and Cartesian forms of the fourth roots are as given below.

z_k	Polar form	Cartesian form
z_0	$2(\cos \pi/6 + i \sin \pi/6)$	$\sqrt{3} + i$
z_1	$2(\cos 2\pi/3 + i \sin 2\pi/3)$	$-1 + \sqrt{3}i$
z_2	$2(\cos 7\pi/6 + i \sin 7\pi/6)$	$-\sqrt{3} - i$
z_3	$2(\cos 5\pi/3 + i \sin 5\pi/3)$	$1 - \sqrt{3}i$

The fourth roots are plotted in Figure 3.7.

Since the principal argument of $-8 + 8\sqrt{3}i$ is $2\pi/3$, its principal fourth root is $z_0 = \sqrt{3} + i$. ∎

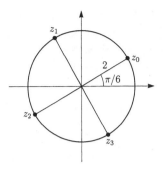

Figure 3.7

The above solution illustrates the following strategy.

Strategy for finding nth roots

To find the n nth roots, $z_0, z_1, \ldots, z_{n-1}$, of a non-zero complex number w:

(a) express w in polar form, with modulus ρ and argument ϕ;

(b) substitute the values of ρ and ϕ in the formula

$$z_k = \rho^{1/n}\left(\cos\left(\frac{\phi}{n} + k\frac{2\pi}{n}\right) + \sin\left(\frac{\phi}{n} + k\frac{2\pi}{n}\right)\right),$$
$$k = 0, 1, \ldots, n-1;$$

(c) if required, convert the roots to Cartesian form.

Remarks

1 In step (a) you should normally choose ϕ to be $\operatorname{Arg} w$ (as in Example 3.2); this has the advantage that the root z_0 obtained in step (b) is the principal nth root of w.

One disadvantage of this choice is the appearance of minus signs when $\operatorname{Arg} w$ is negative. This can be avoided by choosing ϕ to be $(\operatorname{Arg} w) + 2\pi \ (> 0)$, but, with this choice, z_0 in step (b) will not be the principal nth root of w, which has to be identified separately (see the solution to Problem 3.2(b)).

2 Where possible you should try to use the fact that the nth roots of w form a regular n-sided polygon to *check* your calculation of nth roots. For example, in Example 3.2 note that

$$z_1 = iz_0, \quad z_2 = i^2 z_0 = -z_0 \quad \text{and} \quad z_3 = i^3 z_0 = -iz_0,$$

corresponding to the fact that multiplying z by i rotates z about 0 through $\pi/2$ anticlockwise.

Problem 3.2 _____

(a) Determine the cube roots of $8i$ in Cartesian form, plot them in the complex plane, and indicate the principal cube root.

(b) Determine the sixth roots of $-i$ in polar form, plot them in the complex plane, and indicate the principal sixth root.

Problem 3.3 _____

(a) Use the Geometric Series Identity to prove that if z is an nth root of unity $(n \geq 2)$ and $z \neq 1$, then

$$1 + z + z^2 + \cdots + z^{n-1} = 0.$$

(b) Deduce from part (a) that the n nth roots of unity have sum 0.

3.2 Solutions of polynomial equations

The quadratic equation

$$az^2 + bz + c = 0,$$

where a, b, c are complex numbers and $a \neq 0$, can be solved by the methods which are available in the real case. For example, we may be able to factorize the quadratic expression, as in the following cases:

$$z^2 + 9 = (z - 3i)(z + 3i) = 0, \qquad \text{so that } z = \pm 3i;$$
$$z^2 + (1 - i)z - i = (z + 1)(z - i) = 0, \qquad \text{so that } z = -1, i.$$

If there is no easy factorization, then the formula

$$z = \frac{-b \pm \sqrt{b^2 - 4ac}}{2a} \tag{3.2}$$

can be used. The justification of this formula (by completing the square and rearranging) is identical to the real case.

Problem 3.4

Solve the following equations.

(a) $z^2 - 7iz + 8 = 0$ \qquad (b) $z^2 + 2z + 1 - i = 0$

In the previous subsection we saw how to find the n solutions of the equation

$$z^n - w = 0.$$

However, it is only in exceptional cases that we can find an explicit algebraic solution of the polynomial equation (of degree n)

$$a_n z^n + a_{n-1} z^{n-1} + \cdots + a_1 z + a_0 = 0,$$

where $a_0, a_1, \ldots, a_n$ are complex numbers and $a_n \neq 0$. For example, it may be possible to reduce a given polynomial equation to a quadratic equation by making a substitution (as in the next example).

Example 3.3

Solve the equation

$$z^4 + 4z^2 + 8 = 0.$$

Solution

Substituting $w = z^2$ gives

$$w^2 + 4w + 8 = 0,$$

which has solutions

$$w = \frac{-4 \pm \sqrt{16 - 32}}{2} = -2 \pm 2i.$$

Thus $z = \pm\sqrt{-2 + 2i}$ or $z = \pm\sqrt{-2 - 2i}$. Since

$$-2 + 2i = \sqrt{8}(\cos 3\pi/4 + i \sin 3\pi/4),$$

we have

$$\sqrt{-2 + 2i} = 8^{1/4}(\cos 3\pi/8 + i \sin 3\pi/8);$$

thus two solutions of $z^4 + 4z^2 + 8 = 0$ are $\pm 8^{1/4}(\cos 3\pi/8 + i \sin 3\pi/8)$.

Similarly, since

$$-2 - 2i = \sqrt{8}\left(\cos\left(-3\pi/4\right) + i \sin\left(-3\pi/4\right)\right),$$

we have

$$\sqrt{-2 - 2i} = 8^{1/4}\left(\cos(-3\pi/8) + i \sin(-3\pi/8)\right);$$

thus two further solutions are $\pm 8^{1/4}(\cos(-3\pi/8) + i \sin(-3\pi/8))$.

Since $3\pi/4$ is the principal argument of $-2 + 2i$, this is the principal square root of $-2 + 2i$.

So the four solutions are

$$z = \pm 8^{1/4}(\cos 3\pi/8 + i\sin 3\pi/8), \ \pm 8^{1/4}(\cos(-3\pi/8) + i\sin(-3\pi/8)). \quad \blacksquare$$

Remark Since $\cos(-3\pi/8) = \cos 3\pi/8$ and $\sin(-3\pi/8) = -\sin 3\pi/8$, the four solutions in Example 3.3 form two complex conjugate pairs. It can be shown that non-real roots of a polynomial equation with real coefficients must occur in complex conjugate pairs; see Exercise 3.4.

Problem 3.5 _____

(a) Solve the equation

$$z^6 - 7iz^3 + 8 = 0.$$

(*Hint*: Use Problems 3.2(a) and 3.4(a). Also, you may find the following fact useful: if z is a cube root of $8i$, then $-\frac{1}{2}z$ is a cube root of $-i$.)

(b) Solve the equation

$$z^4 + 4iz^2 + 8 = 0.$$

4 SETS OF COMPLEX NUMBERS

After working through this section, you should be able to:

(a) understand the meanings of inequalities between real expressions involving complex numbers;

(b) understand the specification of subsets of the complex plane in terms of such inequalities;

(c) recognize certain basic *open* and *closed sets*.

4.1 Inequalities

Throughout the course we shall use many inequalities involving complex numbers, and you will need to become adept at interpreting them. Here are some simple inequalities involving a complex number z and some examples of values of z for which they are true ($\checkmark$) or false ($\times$).

	$1 + i$	$2 - i$	$-\frac{1}{2} + \frac{1}{2}i$	$-1 - 3i$
$\operatorname{Re} z > 1$	$\times$	$\checkmark$	$\times$	$\times$
$|z| \leq 1$	$\times$	$\times$	$\checkmark$	$\times$
$|\operatorname{Im} z| > 2$	$\times$	$\times$	$\times$	$\checkmark$
$\operatorname{Arg} z < \pi/2$	$\checkmark$	$\checkmark$	$\times$	$\checkmark$

Notice that these inequalities are all between expressions which are *real-valued*. We never write inequalities between complex-valued expressions such as $2 + i$ or $z^2 + 1$.

The inequalities

$$z_1 < z_2 \quad \text{and} \quad z_1 \leq z_2$$

have no meaning unless both z_1 and z_2 are real.

$\mathbb{R}$ is an *ordered* field, but $\mathbb{C}$ is not.

Problem 4.1 _____

Complete the following true/false table.

	$1 + 2i$	$-1 - 2i$	i	-2
Re $z < 0$				
$|z| > 2$				
Im $z \leq -1$				
Arg $z \geq 0$				

4.2 Sketching subsets of the complex plane (audio-tape)

In real analysis we often use intervals, and these are defined by using inequalities. For example, the *open interval* with endpoints 1, 3 is

Some texts use 'round brackets' for open intervals.

$$]1,3[= \{x : 1 < x < 3\}$$

and the *closed interval* with endpoints -2, 2 is

$$[-2,2] = \{x : -2 \leq x \leq 2\}.$$

An interval such as

$$]-\pi, \pi] = \{x : -\pi < x \leq \pi\}$$

is called *half-open* (or *half-closed*), and it is often convenient to use unbounded open and closed intervals, such as

$$]0, \infty[= \{x : x > 0\} \qquad \text{(open)}$$

and

$$[1, \infty[= \{x : x \geq 1\} \qquad \text{(closed)}.$$

In the audio-tape section, which follows, a similar method is used to define various subsets of the complex plane.

NOW START THE TAPE.

1. Line

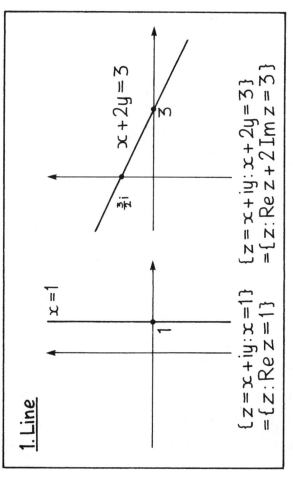

$\{z = x+iy : x = 1\}$
$= \{z : \mathrm{Re}\,z = 1\}$

$\{z = x+iy : x+2y = 3\}$
$= \{z : \mathrm{Re}\,z + 2\,\mathrm{Im}\,z = 3\}$

2. Half-plane

boundary
excluded:
open

boundary
included:
closed

$\{z : \mathrm{Re}\,z > 1\}$

$\{z : \mathrm{Im}\,z \leq 2\}$

3. General half-plane

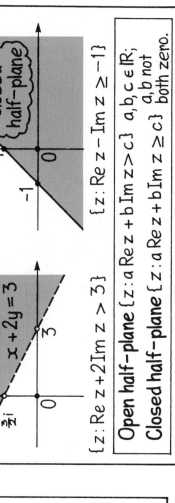

$\{z : \mathrm{Re}\,z + 2\,\mathrm{Im}\,z > 3\}$

open
half-plane
$x+2y = 3$

closed
half-plane
$x-y = -1$

$\{z : \mathrm{Re}\,z - \mathrm{Im}\,z \geq -1\}$

Open half-plane $\{z : a\,\mathrm{Re}\,z + b\,\mathrm{Im}\,z > c\}$ $a, b, c \in \mathbb{R}$;
Closed half-plane $\{z : a\,\mathrm{Re}\,z + b\,\mathrm{Im}\,z \geq c\}$ a, b not both zero.

4. Problem 4.2

(a) Sketch the following sets.
(i) $\{z : 2\,\mathrm{Re}\,z - 3\,\mathrm{Im}\,z = -1\}$ (ii) $\{z : \mathrm{Re}\,z - \mathrm{Im}\,z > 0\}$
(iii) $\{z : \mathrm{Re}\,z + \mathrm{Im}\,z \leq -1\}$
(b) Complete the definition of the following set.

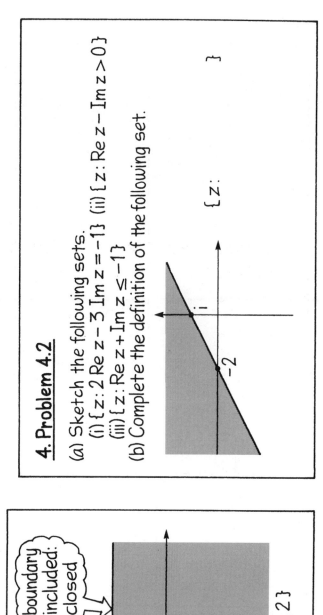

$\{z : \qquad\qquad \}$

5. Circle

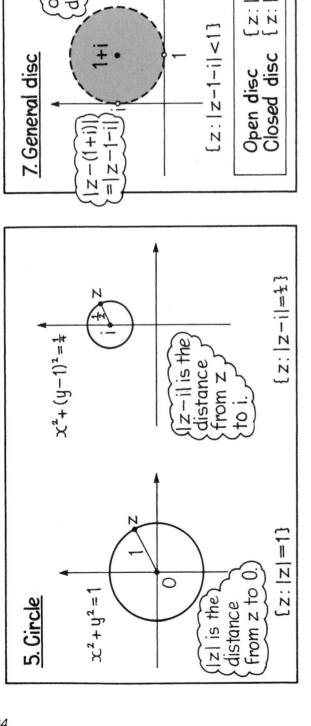

$x^2 + y^2 = 1$

$|z|$ is the distance from z to 0.

$\{z : |z| = 1\}$

$x^2 + (y-1)^2 = \tfrac{1}{4}$

$|z-i|$ is the distance from z to i.

$\{z : |z-i| = \tfrac{1}{2}\}$

7. General disc

$|z-(1+i)| = |z-1-i|$

open disc

$\{z : |z-1-i| < 1\}$

closed disc

$\{z : |z-i| \leq \tfrac{1}{4}\}$

Open disc	$\{z : |z-\alpha| < r\}$	centre $\alpha \in \mathbb{C}$,
Closed disc	$\{z : |z-\alpha| \leq r\}$	radius $r > 0$.

6. Disc

boundary excluded: open

$|z| = 1$

$\{z : |z| < 1\}$

boundary included: closed

$|z| = 2$

$\{z : |z| \leq 2\}$

8. Problem 4.3

(a) Sketch the following sets.
(i) $\{z : |z-1+2i| = 1\}$ (ii) $\{z : |z-1+2i| < 1\}$
(iii) $\{z : |z+2-3i| \leq 3\}$

(b) Complete the definition of the following set (a disc with centre $-1-i$).

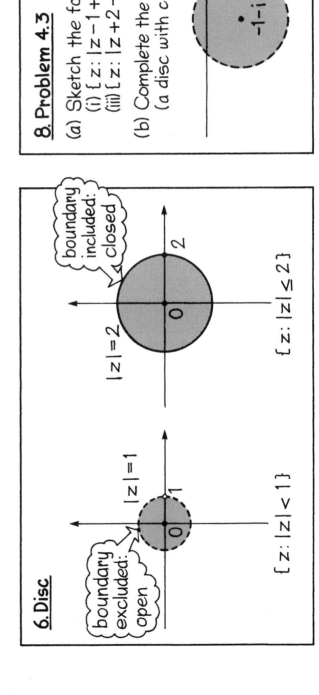

$\{z : \qquad\qquad \}$

11. Punctured disc

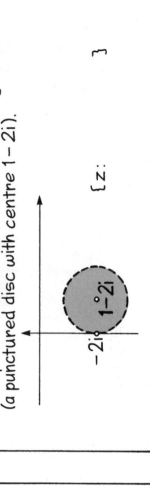

$|z| = 1$

$\{z : 0 < |z| < 1\}$ *(punctured open disc)*

$|z + i| = \tfrac{1}{4}$

$\{z : 0 < |z + i| \leq \tfrac{1}{4}\}$ *(punctured closed disc)*

12. Problem 4.4

(a) Sketch the following sets.

  (i) $\{z : |z + i| > \tfrac{1}{3}\}$ (ii) $\{z : \tfrac{1}{4} \leq |z + 1| < 2\}$

  (iii) $\{z : 2 \leq |z + 2 - 3i| \leq 3\}$

(b) Complete the definition of the following set (a punctured disc with centre $1 - 2i$).

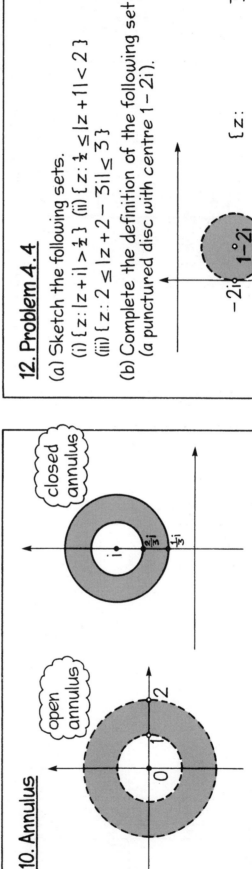

$\{z : \qquad \}$

9. Outside of a disc

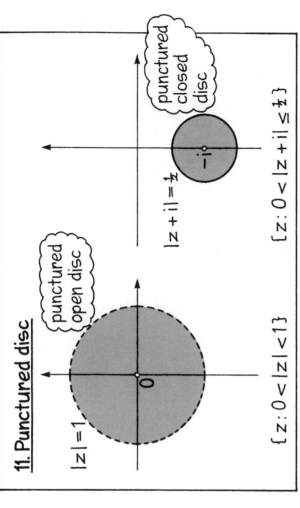

open

$\{z : |z| > 1\}$

closed

$\{z : |z - 1 - i| \geq 1\}$

10. Annulus

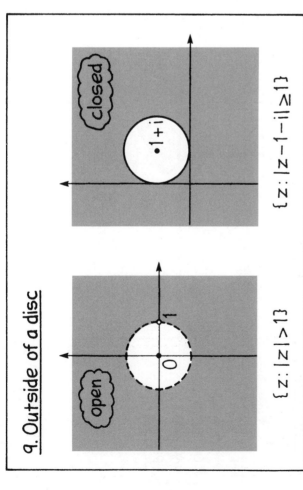

open annulus

$\{z : 1 < |z| < 2\}$

closed annulus

$\{z : \tfrac{1}{3} \leq |z - i| \leq \tfrac{2}{3}\}$

15. General sector

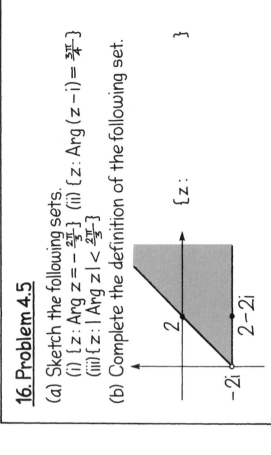

$\{z : |\operatorname{Arg}(z-1)| < \frac{3\pi}{4}\}$

$\{z : -\frac{\pi}{2} \leq \operatorname{Arg}(z-1-i) \leq \frac{\pi}{4}\}$

open sector

$$\boxed{\text{Open sector } \{z : a < \operatorname{Arg}(z-\alpha) < b\} \quad \begin{array}{l} \alpha \in \mathbb{C} \\ -\pi \leq a < b \leq \pi \end{array}}$$

16. Problem 4.5

(a) Sketch the following sets.
  (i) $\{z : \operatorname{Arg} z = -\frac{2\pi}{3}\}$ (ii) $\{z : \operatorname{Arg}(z-i) = \frac{3\pi}{4}\}$
  (iii) $\{z : |\operatorname{Arg} z| < \frac{2\pi}{3}\}$

(b) Complete the definition of the following set.

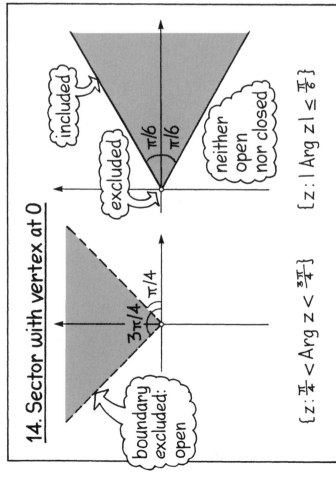

$\{z : \qquad\qquad\}$

13. Ray (half-line)

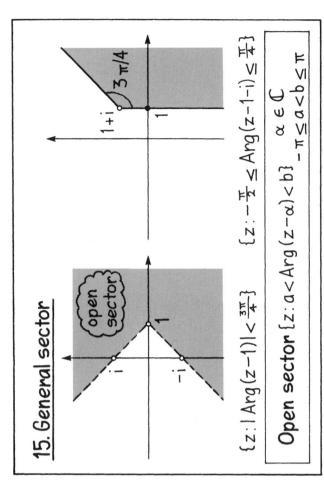

$z - (1+i)$
$= z - 1 - i$

$\{z : \operatorname{Arg} z = \pi/4\}$

$\{z : \operatorname{Arg}(z-1-i) = \pi/4\}$

14. Sector with vertex at 0

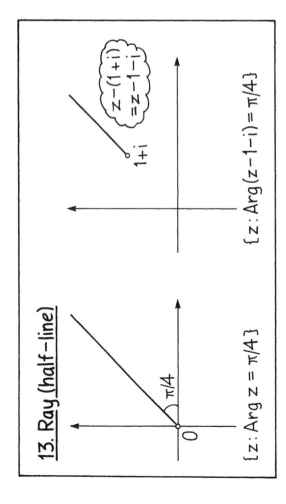

included

excluded

neither open nor closed

boundary excluded: open

$\{z : \frac{\pi}{4} < \operatorname{Arg} z < \frac{3\pi}{4}\}$

$\{z : |\operatorname{Arg} z| \leq \frac{\pi}{6}\}$

17. Operations on sets

Union:
$$A \cup B = \{z : z \in A \text{ or } z \in B\}$$

Intersection:
$$A \cap B = \{z : z \in A \text{ and } z \in B\}$$

Difference:
$$A - B = \{z : z \in A \text{ and } z \notin B\}$$

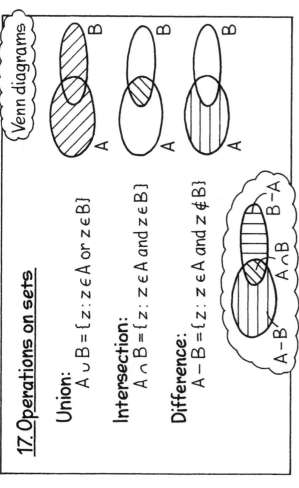

Venn diagrams

19. Complements

$$\mathbb{C} - A = \{z : z \notin A\}$$

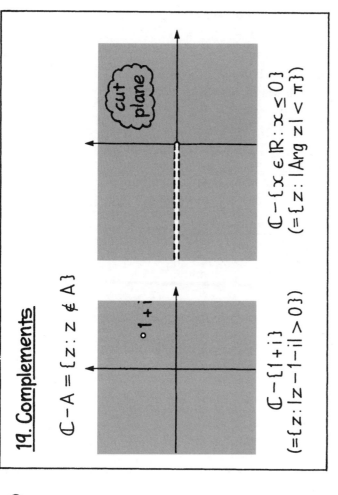

$$\mathbb{C} - \{1+i\}$$
$$(= \{z : |z-1-i| > 0\})$$

cut plane

$$\mathbb{C} - \{x \in \mathbb{R} : x \leq 0\}$$
$$(= \{z : |\operatorname{Arg} z| < \pi\})$$

18. Examples

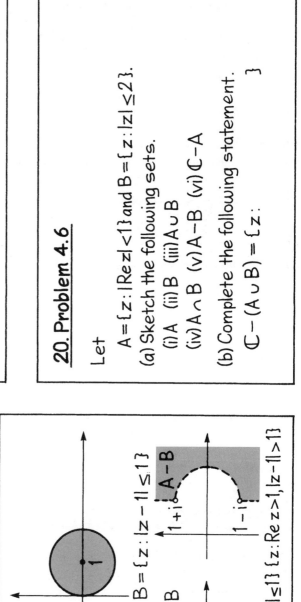

$$A = \{z : \operatorname{Re} z > 1\}$$

$$B = \{z : |z-1| \leq 1\}$$

$A \cap B$

$A - B$

$A \cup B$

$\{z : \operatorname{Re} z > 1 \text{ or } |z-1| \leq 1\}$ $\{z : \operatorname{Re} z > 1, |z-1| \leq 1\}$ $\{z : \operatorname{Re} z > 1, |z-1| > 1\}$

20. Problem 4.6

Let
$$A = \{z : |\operatorname{Re} z| < 1\} \text{ and } B = \{z : |z| \leq 2\}.$$

(a) Sketch the following sets.
   (i) A (ii) B (iii) $A \cup B$
   (iv) $A \cap B$ (v) $A - B$ (vi) $\mathbb{C} - A$

(b) Complete the following statement.
$$\mathbb{C} - (A \cup B) = \{z : \qquad \}$$

In the audio-tape frames we used various conventions for sketching subsets of $\mathbb{C}$:

- the interior of a set is shown by 'tone';
- boundary curves which belong to the set are drawn unbroken;
- boundary curves which do not belong to the set are drawn broken;
- isolated boundary points which belong to the set are drawn as filled-in circles;
- isolated boundary points which do not belong to the set are drawn as empty circles.

These conventions will remain in force throughout the course, although we shall not always include the tone. When sketching sets by hand you should replace tone by hatching.

Problem 4.7 _____

Sketch the following sets, using the conventions listed above.

(a) $\{z : \operatorname{Im} z > 0\}$

(b) $\{z : |z + 1| \leq 1\}$

(c) $\{z : 0 < |z + 1 + 2i| < 1\}$

(d) $\{z : |\operatorname{Arg}(z + 1 - i)| < \pi/3\}$

(e) $\{z : |z - 1| \leq |z - 2|\}$

    (*Hint*: Interpret the inequality in terms of distances.)

(f) $\mathbb{C} - \{z : \operatorname{Re} z \geq 1\}$

(g) $\{z : \operatorname{Im} z > 0\} - \{z : |z + 1| \leq 1\}$

(h) $\{z : \operatorname{Arg} z = \pi/6\} \cup \{z : \operatorname{Arg}(z - \sqrt{3} - i) = 0\}$

(i) $\{z : \operatorname{Arg} z = \pi/6\} \cap \{z : \operatorname{Arg}(z - \sqrt{3} - i) = 0\}$

5 PROVING INEQUALITIES

After working through this section, you should be able to:

(a) use the rules for rearranging inequalities and the rules for obtaining new inequalities from old ones;

(b) prove inequalities involving the moduli of complex numbers by using various forms of the Triangle Inequality.

5.1 Rules for rearranging inequalities

In Section 4 we used equalities and inequalities to define subsets of the complex plane. In this section we show you how to prove new inequalities by deducing them from simpler known inequalities (such as $|z| \geq 0$, which holds for all z) using various rules. We begin by reminding you of the rules for rearranging a given inequality into an *equivalent* form; such equivalent inequalities are linked by the symbol '$\Longleftrightarrow$', which may be read as 'is equivalent to' or 'if and only if'.

Rules for rearranging inequalities

For all a, b, c in $\mathbb{R}$, the following rules apply.

Examples

Rule 1 $a < b \iff b - a > 0$. $x < 3 \iff 3 - x > 0$

Rule 2 $a < b \iff a + c < b + c$. $x < 3 \iff x - 1 < 3 - 1$

Rule 3   If $c > 0$, then $a < b \iff ac < bc$. $2 < x \iff 1 < x/2$

           If $c < 0$, then $a < b \iff ac > bc$. $2 < 3 \iff -2 > -3$

Rule 4   If $a, b > 0$, then $a < b \iff \dfrac{1}{a} > \dfrac{1}{b}$. $2 < 3 \iff \frac{1}{2} > \frac{1}{3}$

Rule 5   If $a, b \geq 0$ and $p > 0$, then $a < b \iff a^p < b^p$. $2 < 3 \iff \sqrt{2} < \sqrt{3}$

Rule 6 $|a| < b \iff -b < a < b$. $|-2| < 3 \iff -3 < -2 < 3$

There are corresponding versions of Rules 1–6 in which the strict inequality '<' is replaced by the weak inequality '≤'.

The next two rules can be used to deduce new inequalities from given ones. Here, however, the new inequalities are not equivalent to the old ones, since the old inequalities cannot be deduced from the new ones. Such deductions are written using the symbol '$\implies$', which may be read as 'implies'.

Transitive Rule

For all a, b, c in $\mathbb{R}$,

$$a < b \text{ and } b < c \implies a < c.$$

For example, if $x < 2$, then $x < 3$ (because $2 < 3$).

Combination Rules

For all a, b, c, d in $\mathbb{R}$, if $a < b$ and $c < d$, then

Sum Rule $a + c < b + d$;

Product Rule $ac < bd$ (provided that $a, c \geq 0$).

For example, if $n < 5$, then as $2 < 3$,
$$n + 2 < 5 + 3 = 8$$
and
$$2n < 3 \times 5 = 15.$$

There are also weak and weak/strict versions of the Transitive Rule and the Combination Rules, which you should be able to work out as they arise.

The following example illustrates how the various rules are used in practice.

Example 5.1

Prove that

$$2r^2 > (r + 1)^2, \qquad \text{for } r \geq 3.$$

Solution

We rearrange the given inequality in order to find an equivalent, but simpler one:

$$2r^2 > (r + 1)^2 \iff 2 > \left(\frac{r + 1}{r}\right)^2 \qquad \text{(Rule 3)}$$

$$\iff \sqrt{2} > 1 + \frac{1}{r} \qquad \text{(Rule 5)}$$

$$\iff \sqrt{2} - 1 > \frac{1}{r} \qquad \text{(Rule 2)}$$

$$\iff r > \frac{1}{\sqrt{2} - 1} = \sqrt{2} + 1 \qquad \text{(Rule 4)}.$$

In future we shall not usually indicate which rule for rearranging a given inequality is being used.

39

Since $\sqrt{2} + 1 = 2.414\ldots$, the final inequality *is* true for $r \geq 3$ (by the Transitive Rule: $r \geq 3$ and $3 > \sqrt{2} + 1 \Longrightarrow r > \sqrt{2} + 1$). Hence the first inequality must be true for $r \geq 3$ also. ∎

Remark Example 5.1 could be solved, alternatively, by using Rule 1 to obtain $r^2 - 2r - 1 > 0$ and then completing the square. There is often more than one way to deal with a given inequality.

Problem 5.1 _____

Prove that
$$\frac{3r}{r^2 + 2} < 1, \qquad \text{for } r > 2.$$

5.2 The Triangle Inequality

Many inequalities have a geometric interpretation. For example, the two inequalities

$$|x| \leq \sqrt{x^2 + y^2} \quad \text{and} \quad |y| \leq \sqrt{x^2 + y^2}$$

state that, in a right-angled triangle, the hypotenuse is the longest side. They can be written in complex form as

$$|\operatorname{Re} z| \leq |z| \quad \text{and} \quad |\operatorname{Im} z| \leq |z| \tag{5.1}$$

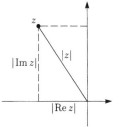

Figure 5.1

(see Figure 5.1) or, equivalently, as

$$-|z| \leq \operatorname{Re} z \leq |z| \quad \text{and} \quad -|z| \leq \operatorname{Im} z \leq |z|. \tag{5.2}$$

Another elementary fact from plane geometry is that the length of any side of a triangle is less than or equal to the sum of the lengths of the other two sides. If z_1, z_2 are complex numbers, then 0, z_1 and $z_1 + z_2$ form the vertices of a triangle (see Figure 5.2) with sides $|z_1|$, $|z_2|$ and $|z_1 + z_2|$, and so

$$|z_1 + z_2| \leq |z_1| + |z_2|.$$

This is one form of an inequality called the Triangle Inequality, which will be used frequently throughout the course.

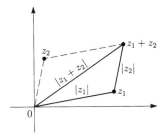

Figure 5.2

Theorem 5.1 Triangle Inequality

If $z_1, z_2 \in \mathbb{C}$, then
(a) $|z_1 + z_2| \leq |z_1| + |z_2|$ (usual form);
(b) $|z_1 - z_2| \geq \big||z_1| - |z_2|\big|$ (backwards form).

Part (b) gives
$$|z_1 - z_2| \geq |z_1| - |z_2|$$
and
$$|z_1 - z_2| \geq |z_2| - |z_1|.$$

Proof Although part (a) follows from plane geometry, we give a proof using complex numbers which illustrates the use of several results from this unit. An alternative proof using polar form is given in Exercise 5.4.

We have

$$
\begin{aligned}
|z_1 + z_2|^2 &= (z_1 + z_2)\overline{(z_1 + z_2)} && \text{(Theorem 2.1(c))} \\
&= (z_1 + z_2)(\overline{z_1} + \overline{z_2}) && \text{(Theorem 1.1(b)(i))} \\
&= z_1\overline{z_1} + z_1\overline{z_2} + z_2\overline{z_1} + z_2\overline{z_2} \\
&= |z_1|^2 + z_1\overline{z_2} + \overline{z_1\overline{z_2}} + |z_2|^2 \\
&\qquad\qquad \text{(Theorem 2.1(c); Theorem 1.1, parts (b)(iii) and (a)(iii))} \\
&= |z_1|^2 + 2\operatorname{Re}(z_1\overline{z_2}) + |z_2|^2 && \text{(Theorem 1.1(a)(i))} \\
&\leq |z_1|^2 + 2|z_1\overline{z_2}| + |z_2|^2 && \text{(by Equation (5.1))} \\
&= |z_1|^2 + 2|z_1||z_2| + |z_2|^2 && \text{(since } |z_1\overline{z_2}| = |z_1||\overline{z_2}| = |z_1||z_2|) \\
&= (|z_1| + |z_2|)^2,
\end{aligned}
$$

and so part (a) follows.

Part (b) can be proved by a similar method; alternatively, note that

$$
\begin{aligned}
|z_1| &= |z_1 - z_2 + z_2| \\
&\leq |z_1 - z_2| + |z_2| && \text{(by part (a))},
\end{aligned}
$$

so that

$$|z_1 - z_2| \geq |z_1| - |z_2|. \tag{5.3}$$

Similarly,

$$|z_2 - z_1| \geq |z_2| - |z_1|,$$

but $|z_2 - z_1| = |z_1 - z_2|$, so that

$$|z_1 - z_2| \geq |z_2| - |z_1|. \tag{5.4}$$

Part (b) follows from inequalities (5.3) and (5.4). ∎

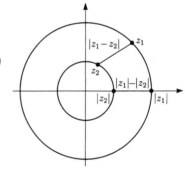

Figure 5.3

The backwards form of the Triangle Inequality also has a useful geometric interpretation, concerning the two circles centred at 0 through z_1 and z_2. It says that the distance from z_1 to z_2 is at least as large as the difference between the radii of these circles, as shown in Figure 5.3 for the case $|z_1| > |z_2|$.

Several other versions of the Triangle Inequality are given in the following corollary. Each is a variant of one of the forms of the Triangle Inequality.

Corollary If $z, z_1, z_2, \ldots, z_n \in \mathbb{C}$, then

(a) $|z| \leq |\operatorname{Re} z| + |\operatorname{Im} z|$;

(b) $|z_1 - z_2| \leq |z_1| + |z_2|$;

(c) $|z_1 + z_2| \geq \big||z_1| - |z_2|\big|$;

(d) $|z_1 \pm z_2 \pm \cdots \pm z_n| \leq |z_1| + |z_2| + \cdots + |z_n|$;

(e) $|z_1 \pm z_2 \pm \cdots \pm z_n| \geq |z_1| - |z_2| - \cdots - |z_n|$.

Parts (a), (b) and (d) are variants of the usual form of the Triangle Inequality, whereas parts (c) and (e) are variants of the backwards form.

Proof Part (a) is obtained by taking $z_1 = \operatorname{Re} z$ and $z_2 = i\operatorname{Im} z$ in the usual form of the Triangle Inequality.

Parts (b) and (c) are obtained by substituting $-z_2$ for z_2 in Theorem 5.1.

Parts (d) and (e) are obtained from Theorem 5.1 and parts (b) and (c) of this corollary by applying the Principle of Mathematical Induction — we omit the details. ∎

The Triangle Inequality can be used to obtain estimates for the modulus of a complex expression involving z when we know that z lies in a certain set (such as a circle). Here are some typical applications.

Example 5.2

(a) Prove that

    (i) $|z^2 - 4z - 3| \leq 15$,   for $|z| = 2$;

    (ii) $|z^2 - 7| \geq 3$,   for $|z| = 2$;

    (iii) $|z^2 + 2| \geq 2$,   for $|z| = 2$.

(b) Find a number M such that

$$\left| \frac{z^2 - 4z - 3}{(z^2 - 7)(z^2 + 2)} \right| \leq M, \qquad \text{for } |z| = 2.$$

Solution

(a) (i) By the Triangle Inequality,

$$|z^2 - 4z - 3| \leq |z^2| + |-4z| + |-3|$$
$$= |z|^2 + 4|z| + 3;$$

so, for $|z| = 2$,

$$|z^2 - 4z - 3| \leq 4 + 8 + 3 = 15.$$

(ii) By the backwards form of the Triangle Inequality,

$$|z^2 - 7| \geq \left| |z|^2 - 7 \right|;$$

so, for $|z| = 2$,

$$|z^2 - 7| \geq |4 - 7| = 3.$$

(iii) By the backwards form of the Triangle Inequality,

$$|z^2 + 2| \geq \left| |z|^2 - 2 \right|;$$

so, for $|z| = 2$,

$$|z^2 + 2| \geq |4 - 2| = 2.$$

As indicated in these solutions it is not usual to refer to any of the variants (in the corollary) of the Triangle Inequality. However, use of the backwards form should be distinguished.

(b) From part (a) we have, for $|z| = 2$,

$$|z^2 - 4z - 3| \leq 15, \quad |z^2 - 7| \geq 3, \quad |z^2 + 2| \geq 2. \qquad (5.5)$$

Now

$$\left| \frac{z^2 - 4z - 3}{(z^2 - 7)(z^2 + 2)} \right| = \frac{|z^2 - 4z - 3|}{|z^2 - 7| \times |z^2 + 2|}$$
$$= |z^2 - 4z - 3| \times \frac{1}{|z^2 - 7|} \times \frac{1}{|z^2 + 2|}.$$

So, for $|z| = 2$, using Inequalities (5.5), we have

$$\left| \frac{z^2 - 4z - 3}{(z^2 - 7)(z^2 + 2)} \right| \leq 15 \times \frac{1}{3} \times \frac{1}{2} = \frac{5}{2},$$

because

$$|z^2 - 7| \geq 3 \implies 1/|z^2 - 7| \leq 1/3$$
$$\text{and } |z^2 + 2| \geq 2 \implies 1/|z^2 + 2| \leq 1/2.$$

Thus we can take $M = 5/2$. ■

Remarks

1 Example 5.2(b) illustrates the fact that to obtain an upper estimate for a quotient, we need an upper estimate (15) for the numerator and a *lower estimate* (3×2) for the denominator.

2 The inequality

$$|z^2 + 2| \geq 2, \qquad \text{for } |z| = 2,$$

is said to be 'best possible' because it holds with equality if $z = 2i$ or $-2i$. $\qquad |(2i)^2 + 2| = |-4 + 2| = 2$

However, the inequality

$$|z^2 - 4z - 3| \leq 15, \qquad \text{for } |z| = 2,$$

is not 'best possible'. With more work it is possible to prove the best possible inequality

$$|z^2 - 4z - 3| \leq 7\sqrt{7/3} \, (= 10.69\ldots), \qquad \text{for } |z| = 2.$$

Problem 5.2

Prove that

(a) $\dfrac{1}{7} \leq \left| \dfrac{1}{3 + 4z^2} \right| \leq 1, \quad \text{for } |z| = 1;$

(b) $2 \leq \left| \dfrac{z^3 + 2z + 1}{z^2 + 1} \right| \leq \dfrac{17}{4}, \quad \text{for } |z| = 3.$

EXERCISES

Section 1

Exercise 1.1 Complete the following table.

z	$\operatorname{Re} z$	$\operatorname{Im} z$	$-z$	$\overline{z}$
$2 + 3i$				
$-3 - i$				
$4i$				
5				
0				

Exercise 1.2 Express each of the following complex numbers in Cartesian form.

(a) i^3 (b) i^4 (c) $(1 + i)^2$ (d) $(1 - i)^2$ (e) $\dfrac{1}{1 - i}$

(f) $\dfrac{1 + i}{1 - i}$ (g) $(1 + i)^4$ (h) $(2 + i)^2 - (2 - i)^2$ (i) $\dfrac{3 + 5i}{2 - 3i}$

(j) $\dfrac{3 + 2i}{1 + 4i}$ (k) $(3 + 4i)^4 - (3 - 4i)^4$ (l) $1 + i + i^2 + \cdots + i^{10}$

(m) $1 - i + i^2 - \cdots + i^{10}$

Exercise 1.3 Write down the real part, imaginary part and complex conjugate of the complex numbers in parts (a), (e), (g) of Exercise 1.2.

Exercise 1.4 Prove that $\operatorname{Im} \overline{z} = -\operatorname{Im} z$.

Section 2

Exercise 2.1 Plot each of the following complex numbers, and express each one in polar form, using the principal argument in each case.

(a) 5 (b) i (c) $-3i$ (d) $2 + 2i$ (e) $-2 + 2i$

(f) $-\sqrt{3} - i$ (g) $3 + 4i$ (h) $3 - 4i$

Exercise 2.2 Plot each of the following complex numbers, and express each one in Cartesian form.

(a) $\cos\pi + i\sin\pi$ (b) $4(\cos(-\pi/2) + i\sin(-\pi/2))$

(c) $3(\cos 3\pi/4 + i\sin 3\pi/4)$ (d) $3(\cos\pi/6 + i\sin\pi/6)$

(e) $\cos(-2\pi/3) + i\sin(-2\pi/3)$

Exercise 2.3 Find the distance from z_2 to z_1 in each of the following cases.

(a) $z_1 = 1 + i$, $z_2 = 2 + 3i$

(b) $z_1 = -2 + 3i$, $z_2 = 1 - 7i$

(c) $z_1 = i$, $z_2 = -i$

Exercise 2.4 Use polar form and de Moivre's Theorem to evaluate the following expressions, giving your answers in Cartesian form.

(a) $(1 + \sqrt{3}i)^5$ (b) $(1 + i)^{-4}$ (c) $\dfrac{(1 + i)^6}{(\sqrt{3} - i)^3}$

Exercise 2.5 Use de Moivre's Theorem and the Binomial Theorem to prove that

$$\sin 3\theta = 3\sin\theta - 4\sin^3\theta,$$

where θ is a real number.

Exercise 2.6 Prove that if $(x + iy)^4 = a + ib$, where $x + iy$ and $a + ib$ are in Cartesian form, then

$$(x^2 + y^2)^4 = a^2 + b^2.$$

Exercise 2.7 Prove that if $\overline{z} = z^{-1}$, then $|z| = 1$.

Section 3

Exercise 3.1 For each of the following complex numbers determine, in Cartesian form where convenient, the nth roots indicated, and plot them. In each case, identify the principal nth root.

(a) The square roots of: (i) $-i$; (ii) $4i$.

(b) The cube roots of: (i) -1; (ii) $-2 + 2i$.

(c) The fourth roots of: (i) $\dfrac{\sqrt{2}}{2}(-1 - i)$; (ii) $-1 + i$.

(d) The fifth roots of: (i) -1; (ii) $-16 + 16\sqrt{3}i$.

Exercise 3.2 Use the method of equating real parts and imaginary parts to solve each of the following equations.

(a) $(x + iy)^2 = 3 + 4i$ (b) $(x + iy)^2 = -5 + 12i$

Exercise 3.3 Solve each of the following equations, and plot their solutions.

(a) $z^4 - z^2 + 1 + i = 0$ (b) $z^3 - 4z^2 + 6z - 4 = 0$

Exercise 3.4 Let $p(z) = a_n z^n + a_{n-1} z^{n-1} + \cdots + a_1 z + a_0$, where $a_0, a_1, \ldots, a_n$ are real numbers. Prove that if z satisfies $p(z) = 0$, then $p(\bar{z}) = 0$. (This shows that non-real roots of a polynomial equation with real coefficients must occur in complex conjugate pairs.)

Section 4

Exercise 4.1 Draw a diagram of each of the following sets of points in the complex plane. In each case indicate which points of the boundary belong to the set, and which points do not.

(a) $\{z : |\operatorname{Re} z| < 1, |\operatorname{Im} z| < 1\}$

(b) $\{z : |z - i| \leq 2, |z| \geq 1\}$

(c) $\{z : \operatorname{Re} z + 2 \operatorname{Im} z + 3 > 0\}$

(d) $\{z : \operatorname{Re} z \geq 0\} \cup \{z : \operatorname{Im} z > 0\}$

(e) $\{z : |z| > 1, |\operatorname{Arg} z| \leq \pi/4\}$

(f) $\{z : -\pi < \operatorname{Arg}(z + 2)\} \cap \{z : \operatorname{Arg}(z + 2) < \pi/2\}$

(g) $\{z : |z + 1 + 2i| \leq 1\}$

(h) $\{z : \operatorname{Re} z > 1, |z - i| < 2\}$

(i) $\{z : |z + i| < |z + 2i|\}$

(j) $\left\{z : \operatorname{Re}\left(\dfrac{z + 1}{z - 1}\right) \leq 0\right\}$

(k) $\{z : |z| < 3\} - \{z : |z| \leq 2\}$

(l) $\mathbb{C} - \{z : z^2 + z - 2 = 0\}$

Section 5

Exercise 5.1 For $|z| = 2$, find an upper estimate for each of the following moduli.

(a) $|z + 3|$ (b) $|z - 4i|$ (c) $|3z + 2|$
(d) $|3z^2 - 5|$ (e) $|z^2 + z + 1|$

Exercise 5.2 For $|z| = 5$, find a positive lower estimate for each of the following moduli.

(a) $|z - 2|$ (b) $|z + 3i|$ (c) $|z - 7|$ (d) $|2z - 7|$

Exercise 5.3 Find (positive) numbers m and M such that

$$m \leq \left|\frac{z^3 + 1}{z^3 - 1}\right| \leq M, \qquad \text{for } |z| = 4.$$

Exercise 5.4 Prove the usual form of the Triangle Inequality:

$$\text{if } z_1, z_2 \in \mathbb{C}, \quad \text{then } |z_1 + z_2| \leq |z_1| + |z_2|,$$

by expressing z_1 and z_2 in polar form.

SOLUTIONS TO THE PROBLEMS

Section 1

1.1 (a)

(i) $(2+i) + 3i(-1+3i) = 2 + i - 3i - 9 = -7 - 2i$

(ii) $(2+i)(-1+3i) = -2 + 6i - i - 3 = -5 + 5i$

(iii) $(-1+3i)(-1-3i) = 1 + 3i - 3i + 9 = 10$

(b) By part (a)(i), $\operatorname{Re} z = -7$ and $\operatorname{Im} z = -2$.

1.2 (a) $(x_1 + iy_1) + (x_2 + iy_2) = (x_1 + x_2) + i(y_1 + y_2)$

(b) $(x_1 + iy_1) - (x_2 + iy_2) = (x_1 - x_2) + i(y_1 - y_2)$

(c) $(x_1 + iy_1)(x_2 + iy_2) = x_1 x_2 + ix_1 y_2 + iy_1 x_2 - y_1 y_2$
$$= (x_1 x_2 - y_1 y_2) + i(x_1 y_2 + y_1 x_2)$$

(d) $(x+iy)(x-iy) = x^2 - ixy + iyx + y^2$
$$= x^2 + y^2$$

1.3 (a) (i) $\dfrac{1}{i} = \dfrac{-i}{i \times (-i)} = \dfrac{-i}{1} = -i$

(ii) $\dfrac{1}{1+i} = \dfrac{1-i}{(1+i)(1-i)} = \dfrac{1-i}{1+1} = \dfrac{1}{2} - \dfrac{1}{2}i$

(iii) $\dfrac{1+2i}{2+3i} = \dfrac{(1+2i)(2-3i)}{(2+3i)(2-3i)} = \dfrac{8+i}{4+9} = \dfrac{8}{13} + \dfrac{1}{13}i$

(b) $\dfrac{x_1 + iy_1}{x_2 + iy_2} = \dfrac{(x_1 + iy_1)(x_2 - iy_2)}{(x_2 + iy_2)(x_2 - iy_2)}$
$$= \dfrac{(x_1 x_2 + y_1 y_2) + i(y_1 x_2 - x_1 y_2)}{x_2^2 + y_2^2}$$
$$= \left(\dfrac{x_1 x_2 + y_1 y_2}{x_2^2 + y_2^2} \right) + i \left(\dfrac{y_1 x_2 - x_1 y_2}{x_2^2 + y_2^2} \right)$$

1.4 *Theorem 1.1(b), part (i)*

Let $z_1 = x_1 + iy_1$, $z_2 = x_2 + iy_2$, so that
$$z_1 + z_2 = (x_1 + x_2) + i(y_1 + y_2).$$
Also $\overline{z_1} = x_1 - iy_1$, $\overline{z_2} = x_2 - iy_2$, so that
$$\overline{z_1} + \overline{z_2} = (x_1 + x_2) - i(y_1 + y_2)$$
$$= \overline{z_1 + z_2}, \quad \text{as required.}$$

Theorem 1.1(b), part (iv)

Let $z_1 = x_1 + iy_1$, $z_2 = x_2 + iy_2$, so that
$$\dfrac{z_1}{z_2} = \left(\dfrac{x_1 x_2 + y_1 y_2}{x_2^2 + y_2^2} \right) + i \left(\dfrac{y_1 x_2 - x_1 y_2}{x_2^2 + y_2^2} \right),$$
by Problem 1.3(b). Also $\overline{z_1} = x_1 - iy_1$, $\overline{z_2} = x_2 - iy_2$, so that
$$\dfrac{\overline{z_1}}{\overline{z_2}} = \left(\dfrac{x_1 x_2 + (-y_1)(-y_2)}{x_2^2 + (-y_2)^2} \right) + i \left(\dfrac{(-y_1)x_2 - x_1(-y_2)}{x_2^2 + (-y_2)^2} \right)$$
$$= \left(\dfrac{x_1 x_2 + y_1 y_2}{x_2^2 + y_2^2} \right) - i \left(\dfrac{y_1 x_2 - x_1 y_2}{x_2^2 + y_2^2} \right)$$
$$= \overline{(z_1/z_2)}, \quad \text{as required.}$$

1.5 (a) $(z_1 + z_2)^3 = (z_1 + z_2)(z_1^2 + 2z_1 z_2 + z_2^2)$
$$= z_1^3 + 2z_1^2 z_2 + z_1 z_2^2 + z_2 z_1^2 + 2z_1 z_2^2 + z_2^3$$
$$= z_1^3 + 3z_1^2 z_2 + 3z_1 z_2^2 + z_2^3$$

(b) $(z_1 - z_2)(z_1^2 + z_1 z_2 + z_2^2)$
$$= z_1^3 + z_1^2 z_2 + z_1 z_2^2 - z_1^2 z_2 - z_1 z_2^2 - z_2^3$$
$$= z_1^3 - z_2^3$$

(c) $(z_1 + z_2)(z_1^2 - z_1 z_2 + z_2^2)$
$$= z_1^3 - z_1^2 z_2 + z_1 z_2^2 + z_1^2 z_2 - z_1 z_2^2 + z_2^3$$
$$= z_1^3 + z_2^3$$

Alternatively, apply part (b) with z_2 replaced by $-z_2$.

1.6 (a) By the Binomial Theorem,
$$(1+i)^4 = 1 + 4i + 6i^2 + 4i^3 + i^4$$
$$= 1 + 4i - 6 - 4i + 1$$
$$= -4.$$

(b) By the Binomial Theorem,
$$(3+2i)^3 = 3^3 + 3 \times 3^2 \times 2i + 3 \times 3 \times (2i)^2 + (2i)^3$$
$$= 27 + 54i - 36 - 8i$$
$$= -9 + 46i.$$

1.7 (a) By the Geometric Series Identity,
$$1 + (1+i) + (1+i)^2 + (1+i)^3$$
$$= \dfrac{1 - (1+i)^4}{1 - (1+i)}$$
$$= \dfrac{1 - (-4)}{-i} \quad \text{(by Problem 1.6(a))}$$
$$= \dfrac{5 \times i}{-i \times i}$$
$$= 5i.$$

(b) By the Geometric Series Identity and the hint,
$$z^5 - i = z^5 - i^5$$
$$= (z - i)(z^4 + z^3 i + z^2 i^2 + zi^3 + i^4)$$
$$= (z - i)(z^4 + z^3 i - z^2 - zi + 1)$$
So $z - 1$ is one linear factor.

Section 2

2.1 With $z_1 = 3 + i$, $z_2 = -1 + 2i$, we have
$$-z_1 = -3 - i, \quad -z_2 = 1 - 2i,$$
$$z_1 + z_2 = 2 + 3i, \quad z_1 - z_2 = 4 - i,$$
$$\overline{z_1} = 3 - i, \quad \overline{z_2} = -1 - 2i,$$
$$\overline{z_1 + z_2} = 2 - 3i.$$

(a)

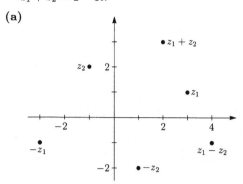

(b)

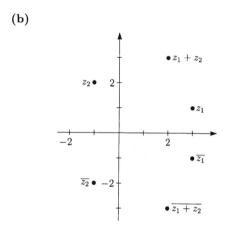

2.2 **(a)** (i) $|1 + i| = \sqrt{1^2 + 1^2} = \sqrt{2}$

(ii) $|2 - 4i| = \sqrt{2^2 + (-4)^2} = \sqrt{20} = 2\sqrt{5}$

(iii) $|i| = \sqrt{1^2} = 1$

(iv) $|-5 + 12i| = \sqrt{(-5)^2 + 12^2} = \sqrt{169} = 13$

(b) If $z = x + iy$, then $\overline{z} = x - iy$ and $-z = -x - iy$. Hence
$$|\overline{z}| = \sqrt{x^2 + (-y)^2} = \sqrt{x^2 + y^2} = |z|,$$
$$|-z| = \sqrt{(-x)^2 + (-y)^2} = \sqrt{x^2 + y^2} = |z|.$$
Note that these results are 'obvious' geometrically.

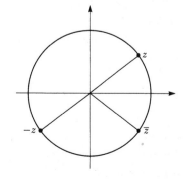

2.3 **(a)** Since $z_1 - z_2 = 4 - i$,
$$|z_1 - z_2| = \sqrt{4^2 + (-1)^2} = \sqrt{17}.$$

(b) Since $z_1 + z_2 = 2 + 3i$,
$$|z_1 + z_2| = \sqrt{2^2 + 3^2} = \sqrt{13}.$$

(c) The distance from z_2 to $-z_1$ is
$$|z_2 - (-z_1)| = |z_2 + z_1| = |z_1 + z_2| = \sqrt{13}.$$

2.4 **(a)** Here $r = |i| = 1$, and the obvious choice for an argument of i is $\theta = \pi/2$. Thus
$$i = 1(\cos \pi/2 + i \sin \pi/2) = \cos \pi/2 + i \sin \pi/2.$$

(b) (i) $2(\cos \pi/3 + i \sin \pi/3) = 2(1/2 + i\sqrt{3}/2)$
$$= 1 + \sqrt{3}i$$

(ii) $3(\cos(-\pi/4) + i \sin(-\pi/4)) = 3(\sqrt{2}/2 + i(-\sqrt{2}/2))$
$$= \frac{3\sqrt{2}}{2} - \frac{3\sqrt{2}}{2}i$$

2.5 In each case we use the strategy to find Arg z.

(a) -4 lies on the negative real axis, so $\text{Arg}(-4) = \pi$ (see Figure 2.12). Since $|-4| = 4$, a polar form of -4 is
$$-4 = 4(\cos \pi + i \sin \pi).$$

(b) $3\sqrt{3} + 3i$ lies in the first quadrant, and
$$\phi = \tan^{-1} \frac{3}{3\sqrt{3}} = \frac{\pi}{6};$$

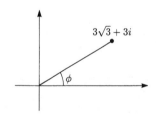

thus the principal argument θ is
$$\theta = \phi \quad \text{(see Figure 2.14)}$$
$$= \pi/6.$$
Since $|3\sqrt{3} + 3i| = \sqrt{(3\sqrt{3})^2 + 3^2} = 6$, a polar form of $3\sqrt{3} + 3i$ is
$$6(\cos \pi/6 + i \sin \pi/6).$$

(c) $\sqrt{3} - i$ lies in the fourth quadrant, and
$$\phi = \tan^{-1} \frac{|-1|}{\sqrt{3}} = \tan^{-1} \frac{1}{\sqrt{3}} = \frac{\pi}{6};$$

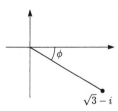

thus the principal argument θ is
$$\theta = -\phi \quad \text{(see Figure 2.14)}$$
$$= -\pi/6.$$
Since $|\sqrt{3} - i| = \sqrt{(\sqrt{3})^2 + (-1)^2} = 2$, a polar form of $\sqrt{3} - i$ is
$$2(\cos(-\pi/6) + i \sin(-\pi/6)).$$

(d) $-1 - i$ lies in the third quadrant, and

$$\phi = \tan^{-1} \frac{|-1|}{|-1|} = \tan^{-1} 1 = \frac{\pi}{4};$$

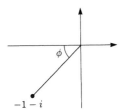

thus the principal argument θ is

$$\theta = -(\pi - \phi) \quad \text{(see Figure 2.14)}$$
$$= -(\pi - \pi/4) = -3\pi/4.$$

Since $|-1 - i| = \sqrt{(-1)^2 + (-1)^2} = \sqrt{2}$, a polar form of $-1 - i$ is

$$\sqrt{2}(\cos(-3\pi/4) + i\sin(-3\pi/4)).$$

2.6 $|z_1| = |-1 - \sqrt{3}i| = 2$ and, from Example 2.2(b), an argument of z_1 is $-2\pi/3$; so

$$z_1 = 2(\cos(-2\pi/3) + i\sin(-2\pi/3)).$$

From Problem 2.5(b),

$$z_2 = 6(\cos\pi/6 + i\sin\pi/6).$$

Thus

$$z_1 z_2 = 2 \times 6(\cos(-2\pi/3 + \pi/6) + i\sin(-2\pi/3 + \pi/6))$$
$$= 12(\cos(-\pi/2) + i\sin(-\pi/2))$$
$$= -12i$$

and

$$z_1^2 = z_1 z_1$$
$$= 2 \times 2(\cos(-2\pi/3 - 2\pi/3) + i\sin(-2\pi/3 - 2\pi/3))$$
$$= 4(\cos(-4\pi/3) + i\sin(-4\pi/3))$$
$$= -2 + 2\sqrt{3}i.$$

2.7 Since $|2i| = 2$ and $\text{Arg}(2i) = \pi/2$, multiplying z by $2i$ scales z by the factor 2 and rotates it anticlockwise through $\pi/2$ about 0.

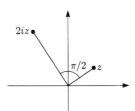

2.8 A polar form of $1 + \sqrt{3}i$ is

$$2(\cos\pi/3 + i\sin\pi/3)$$

and, from Problem 2.5(c), a polar form of $\sqrt{3} - i$ is

$$2(\cos(-\pi/6) + i\sin(-\pi/6)).$$

Thus, from Formula (2.1),

$$z_1/z_2 = \frac{2}{2}\left(\cos(\pi/3 - (-\pi/6)) + i\sin(\pi/3 - (-\pi/6))\right)$$
$$= \cos\pi/2 + i\sin\pi/2$$
$$= i.$$

2.9 Since $|2i| = 2$ and $\text{Arg}(2i) = \pi/2$, dividing z by $2i$ scales z by the factor $\frac{1}{2}$ and rotates it clockwise through $\pi/2$ about 0.

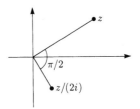

2.10 Since $1 + i = \sqrt{2}(\cos\pi/4 + i\sin\pi/4)$,

$$(1 + i)^{-1} = \frac{1}{\sqrt{2}}(\cos(-\pi/4) + i\sin(-\pi/4))$$
$$= \frac{1}{\sqrt{2}}\left(\frac{1}{\sqrt{2}} - \frac{1}{\sqrt{2}}i\right)$$
$$= \frac{1}{2} - \frac{1}{2}i.$$

2.11 Since

$$z_1 = 1 + i = \sqrt{2}(\cos\pi/4 + i\sin\pi/4),$$
$$z_2 = 1 + \sqrt{3}i = 2(\cos\pi/3 + i\sin\pi/3),$$
$$z_3 = \sqrt{3} + i = 2(\cos\pi/6 + i\sin\pi/6),$$

we have

$$z_1 z_2 z_3 = \sqrt{2} \times 2 \times 2(\cos(\pi/4 + \pi/3 + \pi/6)$$
$$\qquad\qquad + i\sin(\pi/4 + \pi/3 + \pi/6))$$
$$= 4\sqrt{2}(\cos 3\pi/4 + i\sin 3\pi/4)$$
$$= 4\sqrt{2}\left(-\frac{1}{\sqrt{2}} + \frac{1}{\sqrt{2}}i\right)$$
$$= -4 + 4i.$$

2.12 **(a)** Since

$$\sqrt{3} + i = 2(\cos\pi/6 + i\sin\pi/6),$$

de Moivre's Theorem gives

$$(\sqrt{3} + i)^4 = 2^4(\cos 4\pi/6 + i\sin 4\pi/6)$$
$$= 16(\cos 2\pi/3 + i\sin 2\pi/3)$$
$$= 16\left(-\frac{1}{2} + \frac{\sqrt{3}}{2}i\right)$$
$$= -8 + 8\sqrt{3}i.$$

(b) Since

$$1 - \sqrt{3}i = 2(\cos(-\pi/3) + i\sin(-\pi/3)),$$

de Moivre's Theorem gives

$$(1 - \sqrt{3}i)^3 = 2^3(\cos(-3\pi/3) + i\sin(-3\pi/3))$$
$$= 8(\cos(-\pi) + i\sin(-\pi))$$
$$= -8.$$

(c) Since

$$1 + i = \sqrt{2}(\cos\pi/4 + i\sin\pi/4),$$

de Moivre's Theorem gives

$$(1 + i)^{10} = (\sqrt{2})^{10}(\cos 10\pi/4 + i\sin 10\pi/4)$$
$$= 2^5(\cos\pi/2 + i\sin\pi/2)$$
$$= 32i.$$

(d) Since
$$-1 + i = \sqrt{2}(\cos 3\pi/4 + i \sin 3\pi/4),$$
de Moivre's Theorem gives
$$\begin{aligned}
(-1+i)^{-8} &= (\sqrt{2})^{-8}(\cos(-24\pi/4) + i\sin(-24\pi/4)) \\
&= 2^{-4}(\cos(-6\pi) + i\sin(-6\pi)) \\
&= \tfrac{1}{16}.
\end{aligned}$$

(e) Since
$$\sqrt{3} + i = 2(\cos \pi/6 + i \sin \pi/6),$$
de Moivre's Theorem gives
$$\begin{aligned}
(\sqrt{3}+i)^{-6} &= 2^{-6}(\cos(-6\pi/6) + i\sin(-6\pi/6)) \\
&= 2^{-6}(\cos(-\pi) + i\sin(-\pi)) \\
&= -\tfrac{1}{64}.
\end{aligned}$$

Section 3

3.1 Since $-1 + \sqrt{3}i = 2(\cos 2\pi/3 + i \sin 2\pi/3)$, a solution of $z^2 = -1 + \sqrt{3}i$ is obtained by taking z to have modulus $\sqrt{2}$ and argument $\frac{1}{2}(2\pi/3) = \pi/3$.

This gives
$$\begin{aligned}
z &= \sqrt{2}(\cos \pi/3 + i \sin \pi/3) \\
&= \sqrt{2}\left(\frac{1}{2} + \frac{\sqrt{3}}{2}i\right) = \frac{1}{\sqrt{2}} + \frac{\sqrt{3}}{\sqrt{2}}i.
\end{aligned}$$

Hence, by Remark 2, the required solutions are
$$z = \pm\left(\frac{1}{\sqrt{2}} + \frac{\sqrt{3}}{\sqrt{2}}i\right).$$

3.2 **(a)** Using the principal argument for $8i$, we have
$$8i = 8(\cos \pi/2 + i \sin \pi/2),$$
and, using the strategy, we deduce that the cube roots of $8i$ are
$$z_k = 8^{1/3}\left(\cos\left(\frac{\pi}{6} + k\frac{2\pi}{3}\right) + i\sin\left(\frac{\pi}{6} + k\frac{2\pi}{3}\right)\right),$$
$$k = 0, 1, 2;$$
that is,
$$\begin{aligned}
z_0 &= 2(\cos \pi/6 + i \sin \pi/6) = \sqrt{3} + i, \\
z_1 &= 2(\cos 5\pi/6 + i \sin 5\pi/6) = -\sqrt{3} + i, \\
z_2 &= 2(\cos 3\pi/2 + i \sin 3\pi/2) = -2i.
\end{aligned}$$
Since the principal argument of $8i$ is $\pi/2$, the principal cube root of $8i$ is z_0.

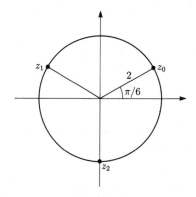

(b) The principal argument of $-i$ is $-\pi/2$; to avoid negative signs, we use the argument $3\pi/2$. Thus
$$-i = \cos 3\pi/2 + i \sin 3\pi/2,$$

and, using the strategy, we deduce that the sixth roots of $-i$ are
$$z_k = \cos\left(\frac{\pi}{4} + k\frac{\pi}{3}\right) + i\sin\left(\frac{\pi}{4} + k\frac{\pi}{3}\right), \quad k = 0, 1, \ldots, 5;$$
that is,
$$\begin{aligned}
z_0 &= \cos \pi/4 + i \sin \pi/4, \\
z_1 &= \cos 7\pi/12 + i \sin 7\pi/12, \\
z_2 &= \cos 11\pi/12 + i \sin 11\pi/12, \\
z_3 &= \cos 15\pi/12 + i \sin 15\pi/12, \\
z_4 &= \cos 19\pi/12 + i \sin 19\pi/12, \\
z_5 &= \cos 23\pi/12 + i \sin 23\pi/12.
\end{aligned}$$
Since the principal argument of $-i$ is $-\pi/2$, the principal sixth root of $-i$ has argument $-\pi/12$ and hence it is z_5.

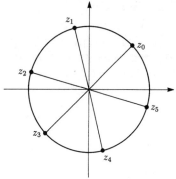

Alternatively, using the principal argument, we have
$$-i = \cos(-\pi/2) + i \sin(-\pi/2),$$
and, using the strategy, we deduce that the sixth roots of $-i$ are
$$z_k = \cos\left(-\frac{\pi}{12} + k\frac{\pi}{3}\right) + i\sin\left(-\frac{\pi}{12} + k\frac{\pi}{3}\right),$$
$$k = 0, 1, \ldots, 5;$$
that is,
$$\begin{aligned}
z_0 &= \cos(-\pi/12) + i \sin(-\pi/12), \\
z_1 &= \cos \pi/4 + i \sin \pi/4, \\
z_2 &= \cos 7\pi/12 + i \sin 7\pi/12, \\
z_3 &= \cos 11\pi/12 + i \sin 11\pi/2, \\
z_4 &= \cos 15\pi/12 + i \sin 15\pi/12, \\
z_5 &= \cos 19\pi/12 + i \sin 19\pi/12.
\end{aligned}$$
The principal sixth root is
$$z_0 = \cos(-\pi/12) + i \sin(-\pi/12).$$

3.3 **(a)** By the Geometric Series Identity,
$$1 - z^n = (1 - z)(1 + z + z^2 + \cdots + z^{n-1}).$$
Thus if
$$z^n = 1 \text{ but } z \neq 1,$$
then
$$1 - z^n = 0 \text{ but } 1 - z \neq 0,$$
so that
$$1 + z + z^2 + \cdots + z^{n-1} = 0,$$
as required.

(b) By the corollary to Theorem 3.1 and de Moivre's Theorem, the nth roots of unity are of the form
$$1, z_1, z_1^2, \ldots, z_1^{n-1},$$
where
$$z_1 = \cos 2\pi/n + i \sin 2\pi/n.$$
Hence, by part (a), the sum of the nth roots of unity is 0.

3.4 **(a)** The factorization

$$z^2 - 7iz + 8 = (z + i)(z - 8i) = 0$$

shows that the solutions are $z = -i,\ 8i$.

(b) Formula (3.2) with $a = 1$, $b = 2$, $c = 1 - i$ gives

$$z = \frac{-2 \pm \sqrt{4 - 4(1 - i)}}{2}$$

$$= -1 \pm \sqrt{i}$$

$$= -1 \pm \frac{1}{\sqrt{2}}(1 + i) \quad \text{(by Example 3.1)}$$

$$= \left(-1 + \frac{1}{\sqrt{2}}\right) + \frac{1}{\sqrt{2}}i,\ \left(-1 - \frac{1}{\sqrt{2}}\right) - \frac{1}{\sqrt{2}}i.$$

3.5 **(a)** Putting $w = z^3$ gives

$$w^2 - 7iw + 8 = 0,$$

and so $w = -i$ or $8i$ (see Problem 3.4(a)). Thus the six solutions are the cube roots of $-i$ and $8i$. We found the cube roots of $8i$ in Problem 3.2 — these are

$$\sqrt{3} + i,\quad -\sqrt{3} + i \quad \text{and} \quad -2i.$$

By a similar method, or using the hint, we find that the cube roots of $-i$ are

$$-\frac{\sqrt{3}}{2} - \frac{1}{2}i,\quad \frac{\sqrt{3}}{2} - \frac{1}{2}i \quad \text{and} \quad i.$$

Hence the six solutions are

$$\sqrt{3} + i,\ -\sqrt{3} + i,\ -2i,\ -\frac{\sqrt{3}}{2} - \frac{1}{2}i,\ \frac{\sqrt{3}}{2} - \frac{1}{2}i,\ i.$$

(b) Putting $w = z^2$ gives

$$w^2 + 4iw + 8 = 0,$$

and so

$$w = \frac{-4i \pm \sqrt{-16 - 32}}{2};$$

that is,

$$w = (\sqrt{12} - 2)i \quad \text{or} \quad w = -(\sqrt{12} + 2)i.$$

Since

$$(\sqrt{12} - 2)i = (\sqrt{12} - 2)(\cos \pi/2 + i \sin \pi/2)$$

and

$$-(\sqrt{12} + 2)i = (\sqrt{12} + 2)(\cos 3\pi/2 + i \sin 3\pi/2),$$

the four solutions are

$$z = \pm(\sqrt{12} - 2)^{1/2}(\cos \pi/4 + i \sin \pi/4)$$

$$= \pm(\sqrt{12} - 2)^{1/2}\left(\frac{1}{\sqrt{2}} + \frac{1}{\sqrt{2}}i\right)$$

$$= \pm(\sqrt{3} - 1)^{1/2}(1 + i)$$

and

$$z = \pm(\sqrt{12} + 2)^{1/2}(\cos 3\pi/4 + i \sin 3\pi/4)$$

$$= \pm(\sqrt{12} + 2)^{1/2}\left(-\frac{1}{\sqrt{2}} + \frac{1}{\sqrt{2}}i\right)$$

$$= \pm(\sqrt{3} + 1)^{1/2}(-1 + i).$$

Section 4

4.1

	$1 + 2i$	$-1 - 2i$	i	-2
$\text{Re}\,z < 0$	×	✓	×	✓
$\lvert z \rvert > 2$	✓	✓	×	×
$\text{Im}\,z \leq -1$	×	✓	×	×
$\text{Arg}\,z \geq 0$	✓	×	✓	✓

4.2 **(a)** (i)

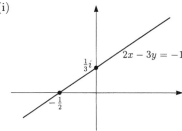

$$\{z : 2\,\text{Re}\,z - 3\,\text{Im}\,z = -1\}$$

(ii)

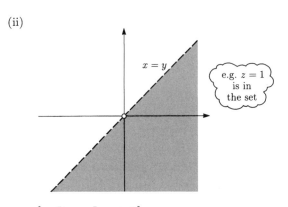

$$\{z : \text{Re}\,z - \text{Im}\,z > 0\}$$

(iii)

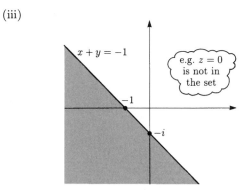

$$\{z : \text{Re}\,z + \text{Im}\,z \leq -1\}$$

(b) The half-plane contains its boundary, which has equation

$$\tfrac{1}{2}x - y = -1.$$

At $z = 0$, $\tfrac{1}{2}x - y > -1$, and since 0 is not in this half-plane, the half-plane is

$$\left\{z : \tfrac{1}{2}\,\text{Re}\,z - \text{Im}\,z \leq -1\right\}.$$

4.3 (a) (i)

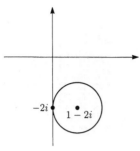

$$\{z : |z - 1 + 2i| = 1\}$$

(ii)

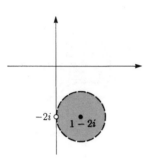

$$\{z : |z - 1 + 2i| < 1\}$$

(iii)

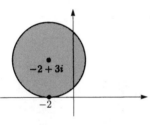

$$\{z : |z + 2 - 3i| \le 3\}$$

(b) This set is an open disc with centre $-1 - i$ and radius $|-1 - i| = \sqrt{2}$, so it is

$$\{z : |z + 1 + i| < \sqrt{2}\}.$$

4.4 (a) (i)

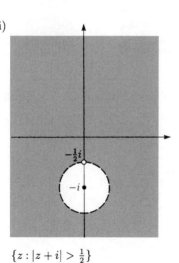

$$\{z : |z + i| > \tfrac{1}{2}\}$$

(ii)

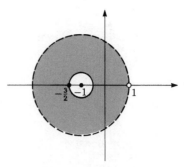

$$\{z : \tfrac{1}{2} \le |z + 1| < 2\}$$

(iii)

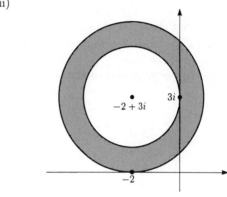

$$\{z : 2 \le |z + 2 - 3i| \le 3\}$$

(b) This set is a punctured open disc with centre $1 - 2i$ and radius 1, so it is

$$\{z : 0 < |z - 1 + 2i| < 1\}.$$

4.5 (a) (i)

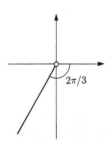

$$\{z : \operatorname{Arg} z = -2\pi/3\}$$

(ii)

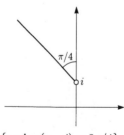

$$\{z : \operatorname{Arg}(z - i) = 3\pi/4\}$$

(iii)

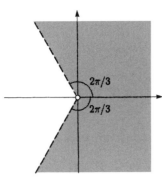

$$\{z : |\operatorname{Arg} z| < 2\pi/3\}$$

(b) This set is a sector (not an open one) having the rays

$$\{z : \operatorname{Arg}(z + 2i) = 0\} \quad \text{and} \quad \{z : \operatorname{Arg}(z + 2i) = \pi/4\}$$

as boundary, so it is

$$\{z : 0 \leq \operatorname{Arg}(z + 2i) \leq \pi/4\}.$$

4.6 (a) (i)

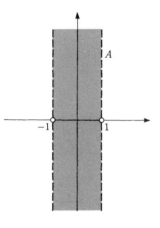

(ii)

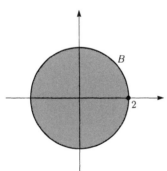

(iii)

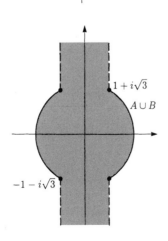

(iv)

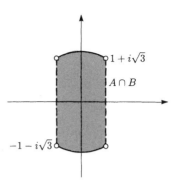

(v)

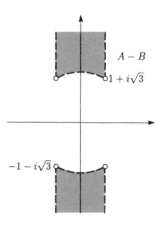

(vi)

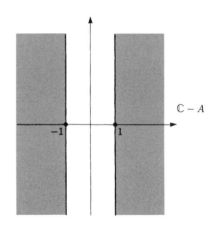

(b) $\mathbb{C} - (A \cup B)$ is the set of points z which lie in neither A nor B; that is,

$$\{z : |\operatorname{Re} z| \geq 1, |z| > 2\}.$$

4.7 **(a)**

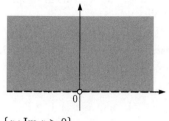

$\{z : \operatorname{Im} z > 0\}$

(b)

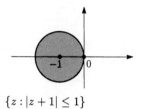

$\{z : |z + 1| \le 1\}$

(c)

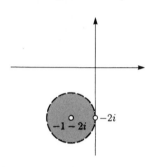

$\{z : 0 < |z + 1 + 2i| < 1\}$

(d)

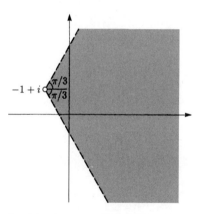

$\{z : |\operatorname{Arg}(z + 1 - i)| < \pi/3\}$

(e)

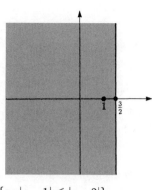

$\{z : |z - 1| \le |z - 2|\}$

This set consists of all points z whose distance from 1 is less than or equal to their distance from 2.

(f)

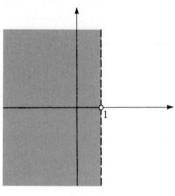

$\mathbb{C} - \{z : \operatorname{Re} z \ge 1\}$

(g)

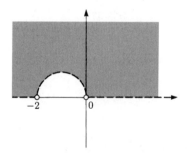

$\{z : \operatorname{Im} z > 0\} - \{z : |z + 1| \le 1\}$

(h)

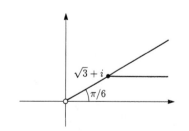

$\{z : \operatorname{Arg} z = \pi/6\} \cup \{z : \operatorname{Arg}(z - \sqrt{3} - i) = 0\}$

(i) The sets $\{z : \operatorname{Arg} z = \pi/6\}$ and $\{z : \operatorname{Arg}(z - \sqrt{3} - i) = 0\}$ have no points in common: their intersection is the empty set, $\varnothing$. We make no attempt to sketch $\varnothing$!

Section 5

5.1 Rearranging the given inequality, we obtain

$$\frac{3r}{r^2 + 2} < 1 \iff 3r < r^2 + 2 \quad (\text{since } r^2 + 2 > 0, \text{ for all } r)$$

$$\iff 0 < r^2 - 3r + 2$$

$$\iff 0 < (r-1)(r-2).$$

Since the final inequality is true for $r > 2$, the first inequality must be true for $r > 2$.

5.2 (a) By the Triangle Inequality,

$$|3 + 4z^2| \leq |3| + |4z^2| = 3 + 4|z|^2.$$

Hence, for $|z| = 1$,

$$|3 + 4z^2| \leq 7,$$

so that

$$\left| \frac{1}{3 + 4z^2} \right| \geq \frac{1}{7}.$$

Now, by the backwards form of the Triangle Inequality,

$$|3 + 4z^2| \geq \left| 4|z|^2 - |3| \right|.$$

Hence, for $|z| = 1$,

$$|3 + 4z^2| \geq |4 - 3| = 1,$$

so that

$$\left| \frac{1}{3 + 4z^2} \right| \leq \frac{1}{1} = 1.$$

Thus

$$\frac{1}{7} \leq \left| \frac{1}{3 + 4z^2} \right| \leq 1, \quad \text{for } |z| = 1,$$

as required.

(b) We first establish the right-hand inequality. By the Triangle Inequality,

$$|z^3 + 2z + 1| \leq |z|^3 + 2|z| + 1,$$

and, by the backwards form of the Triangle Inequality,

$$|z^2 + 1| \geq \left| |z|^2 - 1 \right|.$$

Hence, for $|z| = 3$,

$$|z^3 + 2z + 1| \leq 34 \quad \text{and} \quad |z^2 + 1| \geq 8,$$

so that

$$\left| \frac{z^3 + 2z + 1}{z^2 + 1} \right| \leq \frac{34}{8} = \frac{17}{4},$$

as required.

Next, we establish the left-hand inequality. By the backwards form of the Triangle Inequality,

$$|z^3 + 2z + 1| \geq |z|^3 - 2|z| - 1,$$

and, by the usual form of the Triangle Inequality,

$$|z^2 + 1| \leq |z|^2 + 1.$$

Hence, for $|z| = 3$,

$$|z^3 + 2z + 1| \geq 20 \quad \text{and} \quad |z^2 + 1| \leq 10,$$

so that

$$\left| \frac{z^3 + 2z + 1}{z^2 + 1} \right| \geq \frac{20}{10} = 2,$$

as required.

SOLUTIONS TO THE EXERCISES

Section 1

1.1

z	Re z	Im z	$-z$	$\overline{z}$
$2+3i$	2	3	$-2-3i$	$2-3i$
$-3-i$	-3	-1	$3+i$	$-3+i$
$4i$	0	4	$-4i$	$-4i$
5	5	0	-5	5
0	0	0	0	0

1.2 (a) $i^3 = (i^2)i = -i$

(b) $i^4 = (i^2)(i^2) = 1$

(c) $(1+i)^2 = 1 + 2i + i^2 = 2i$

(d) $(1-i)^2 = 1 - 2i + (-i)^2 = -2i$

(e) $\dfrac{1}{1-i} = \dfrac{1+i}{(1-i)(1+i)} = \dfrac{1+i}{2}$

(f) $\dfrac{1+i}{1-i} = \dfrac{(1+i)(1+i)}{(1-i)(1+i)} = \dfrac{2i}{2} = i$

(g) $(1+i)^4 = ((1+i)^2)^2 = (2i)^2 = -4$
(or use the Binomial Theorem)

(h) $(2+i)^2 - (2-i)^2 = (4+4i-1) - (4-4i-1) = 8i$

(i) $\dfrac{3+5i}{2-3i} = \dfrac{(3+5i)(2+3i)}{(2-3i)(2+3i)}$

$= \dfrac{6+19i-15}{4+9} = -\dfrac{9}{13} + \dfrac{19}{13}i$

(j) $\dfrac{3+2i}{1+4i} = \dfrac{(3+2i)(1-4i)}{(1+4i)(1-4i)}$

$= \dfrac{3-10i+8}{1+16} = \dfrac{11}{17} - \dfrac{10}{17}i$

(k) By the Binomial Theorem,

$(3+4i)^4 - (3-4i)^4$

$= (3^4 + 4 \times 3^3 \times (4i) + 6 \times 3^2 \times (4i)^2$
$\quad + 4 \times 3 \times (4i)^3 + (4i)^4)$
$\quad - (3^4 + 4 \times 3^3 \times (-4i) + 6 \times 3^2 \times (-4i)^2$
$\quad + 4 \times 3 \times (-4i)^3 + (-4i)^4)$

$= 2(432i - 768i) = -672i$

(l) By the Geometric Series Identity,

$1 + i + i^2 + \cdots + i^{10} = \dfrac{1 - i^{11}}{1 - i}$

$= \dfrac{1+i}{1-i} = \tfrac{1}{2}(1+i)^2 = i$

(m) By the Geometric Series Identity,

$1 - i + i^2 - \cdots + i^{10} = \dfrac{1 - (-i)^{11}}{1 - (-i)}$

$= \dfrac{1-i}{1+i}$

$= \dfrac{(1-i)(1-i)}{(1+i)(1-i)} = \dfrac{-2i}{2} = -i$

1.3

Complex number	Real part	Imaginary part	Conjugate
(a) $-i$	0	-1	i
(e) $(1+i)/2$	$\frac{1}{2}$	$\frac{1}{2}$	$(1-i)/2$
(g) -4	-4	0	-4

1.4 If $z = x + iy$, then $\overline{z} = x - iy$ and

$\text{Im}\,\overline{z} = \text{Im}(x - iy)$

$= -y$

$= -\text{Im}(x + iy)$

$= -\text{Im}\,z.$

Section 2

2.1 If you can 'see' what the principal argument is from your diagram, write it down. Our calculation of it uses the strategy in Section 2.

(a)

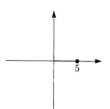

$5 = 5(\cos 0 + i \sin 0)$

(b)

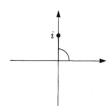

$i = 1\left(\cos\dfrac{\pi}{2} + i\sin\dfrac{\pi}{2}\right)$

(c)

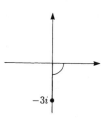

$-3i = 3\left(\cos\left(-\dfrac{\pi}{2}\right) + i\sin\left(-\dfrac{\pi}{2}\right)\right)$

(d)

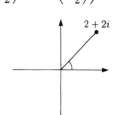

$|2+2i| = \sqrt{8},$

$\text{Arg}(2+2i) = \tan^{-1}\dfrac{2}{2} = \dfrac{\pi}{4},$

$2 + 2i = \sqrt{8}\left(\cos\dfrac{\pi}{4} + i\sin\dfrac{\pi}{4}\right).$

(e)

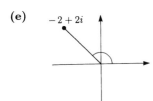

$|-2 + 2i| = \sqrt{8},$
$\text{Arg}(-2 + 2i) = \pi - \tan^{-1}\dfrac{2}{2} = \dfrac{3\pi}{4},$
$-2 + 2i = \sqrt{8}\left(\cos\dfrac{3\pi}{4} + i\sin\dfrac{3\pi}{4}\right).$

(f)

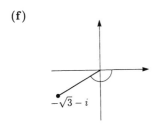

$|-\sqrt{3} - i| = 2,$
$\text{Arg}(-\sqrt{3} - i) = -\left(\pi - \tan^{-1}\dfrac{1}{\sqrt{3}}\right) = -\dfrac{5\pi}{6},$
$-\sqrt{3} - i = 2\left(\cos\left(-\dfrac{5\pi}{6}\right) + i\sin\left(-\dfrac{5\pi}{6}\right)\right).$

(g)

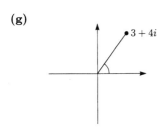

$|3 + 4i| = \sqrt{25} = 5,$
$\text{Arg}(3 + 4i) = \tan^{-1}\frac{4}{3},$
$3 + 4i = 5(\cos\theta + i\sin\theta),$
where $\theta = \tan^{-1}\frac{4}{3} \simeq 0.927\,\text{rad}.$

(h)

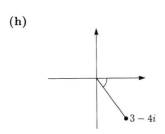

$|3 - 4i| = \sqrt{25} = 5,$
$\text{Arg}(3 - 4i) = -\tan^{-1}\frac{4}{3},$
$3 - 4i = 5(\cos\theta + i\sin\theta),$
where $\theta = -\tan^{-1}\frac{4}{3} \simeq -0.927\,\text{rad}.$

2.2 (a)

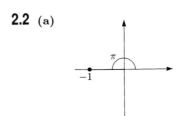

$\cos\pi + i\sin\pi = -1$

(b)

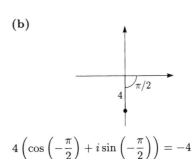

$4\left(\cos\left(-\dfrac{\pi}{2}\right) + i\sin\left(-\dfrac{\pi}{2}\right)\right) = -4i$

(c)

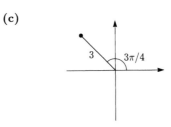

$3\left(\cos\dfrac{3\pi}{4} + i\sin\dfrac{3\pi}{4}\right) = -\dfrac{3\sqrt{2}}{2} + \dfrac{3\sqrt{2}}{2}i$

(d)

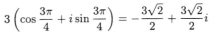

$3\left(\cos\dfrac{\pi}{6} + i\sin\dfrac{\pi}{6}\right) = \dfrac{3\sqrt{3}}{2} + \dfrac{3}{2}i$

(e)

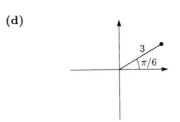

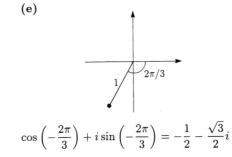

$\cos\left(-\dfrac{2\pi}{3}\right) + i\sin\left(-\dfrac{2\pi}{3}\right) = -\dfrac{1}{2} - \dfrac{\sqrt{3}}{2}i$

2.3 (a) $|z_2 - z_1| = |(2 + 3i) - (1 + i)|$
$$= |1 + 2i|$$
$$= \sqrt{1 + 4} = \sqrt{5}$$

(b) $|z_2 - z_1| = |(1 - 7i) - (-2 + 3i)|$
$$= |3 - 10i|$$
$$= \sqrt{9 + 100} = \sqrt{109}.$$

(c) $|z_2 - z_1| = |-i - i|$
$$= |-2i|$$
$$= 2$$

2.4 (a) $|1 + \sqrt{3}i| = \sqrt{1 + 3} = 2$,
$\text{Arg}(1 + \sqrt{3}i) = \tan^{-1} \dfrac{\sqrt{3}}{1} = \dfrac{\pi}{3};$
so
$$1 + \sqrt{3}i = 2\left(\cos\frac{\pi}{3} + i\sin\frac{\pi}{3}\right).$$
Thus
$$(1 + \sqrt{3}i)^5 = \left(2\left(\cos\frac{\pi}{3} + i\sin\frac{\pi}{3}\right)\right)^5$$
$$= 2^5\left(\cos\frac{5\pi}{3} + i\sin\frac{5\pi}{3}\right),$$
$$\text{by de Moivre's Theorem,}$$
$$= 32\left(\frac{1}{2} - \frac{\sqrt{3}}{2}i\right)$$
$$= 16 - 16\sqrt{3}i.$$

(b) $|1 + i| = \sqrt{2}$, $\text{Arg}(1 + i) = \dfrac{\pi}{4};$
so
$$1 + i = \sqrt{2}\left(\cos\frac{\pi}{4} + i\sin\frac{\pi}{4}\right).$$
Thus
$$(1 + i)^{-4} = \left(\sqrt{2}\left(\cos\frac{\pi}{4} + i\sin\frac{\pi}{4}\right)\right)^{-4}$$
$$= 2^{-4/2}\left(\cos\frac{-4\pi}{4} + i\sin\frac{-4\pi}{4}\right),$$
$$\text{by de Moivre's Theorem,}$$
$$= \tfrac{1}{4}(\cos(-\pi) + i\sin(-\pi))$$
$$= -\tfrac{1}{4}.$$

(c) $1 + i = \sqrt{2}\left(\cos\dfrac{\pi}{4} + i\sin\dfrac{\pi}{4}\right);$
thus
$$(1 + i)^6 = \left(\sqrt{2}\left(\cos\frac{\pi}{4} + i\sin\frac{\pi}{4}\right)\right)^6$$
$$= 2^3\left(\cos\frac{3\pi}{2} + i\sin\frac{3\pi}{2}\right),$$
$$\text{by de Moivre's Theorem.}$$
Also, $|\sqrt{3} - i| = \sqrt{3 + 1} = 2$,
$\text{Arg}\left(\sqrt{3} - i\right) = -\tan^{-1}\dfrac{1}{\sqrt{3}} = -\dfrac{\pi}{6};$
so
$$\sqrt{3} - i = 2\left(\cos\left(-\frac{\pi}{6}\right) + i\sin\left(-\frac{\pi}{6}\right)\right).$$
Thus
$$(\sqrt{3} - i)^{-3} = \left(2\left(\cos\left(-\frac{\pi}{6}\right) + i\sin\left(-\frac{\pi}{6}\right)\right)\right)^{-3}$$
$$= 2^{-3}\left(\cos\frac{\pi}{2} + i\sin\frac{\pi}{2}\right).$$

Hence
$$\frac{(1 + i)^6}{(\sqrt{3} - i)^3}$$
$$= 2^3\left(\cos\frac{3\pi}{2} + i\sin\frac{3\pi}{2}\right) \times 2^{-3}\left(\cos\frac{\pi}{2} + i\sin\frac{\pi}{2}\right)$$
$$= \cos 2\pi + i\sin 2\pi = 1.$$

2.5 By de Moivre's Theorem,
$$(\cos\theta + i\sin\theta)^3 = \cos 3\theta + i\sin 3\theta.$$
By the Binomial Theorem,
$$(\cos\theta + i\sin\theta)^3$$
$$= \cos^3\theta + 3\cos^2\theta(i\sin\theta) + 3\cos\theta(i\sin\theta)^2 + (i\sin\theta)^3$$
$$= \cos^3\theta - 3\cos\theta\sin^2\theta + i(3\cos^2\theta\sin\theta - \sin^3\theta).$$
The two expressions we have obtained for $(\cos\theta + i\sin\theta)^3$ are equal, and so their real parts are equal and their imaginary parts are equal. Equating the two imaginary parts gives
$$\sin 3\theta = 3\cos^2\theta\sin\theta - \sin^3\theta$$
$$= 3\sin\theta - 4\sin^3\theta \quad (\text{since } \cos^2\theta = 1 - \sin^2\theta),$$
as required.

2.6 If $(x + iy)^4 = a + ib$, then
$$|x + iy|^4 = |a + ib|,$$
by Theorem 2.1(e), so that
$$(x^2 + y^2)^4 = a^2 + b^2.$$

2.7 If $\bar{z} = z^{-1}$, then
$$z\bar{z} = zz^{-1} = 1.$$
Hence, by Theorem 2.1(c),
$$|z|^2 = 1,$$
so that $|z| = 1$, as required.

Section 3

3.1 In each case the given complex number is expressed in polar form using the principal argument.

(a) (i) $-i = \cos(-\pi/2) + i\sin(-\pi/2)$; the square roots of $-i$ are
$$z_k = \cos(-\pi/4 + k\pi) + i\sin(-\pi/4 + k\pi), \quad k = 0, 1.$$
Thus the principal square root is
$$z_0 = \frac{\sqrt{2}}{2} - \frac{\sqrt{2}}{2}i = \frac{\sqrt{2}}{2}(1 - i),$$
and the other root is
$$z_1 = -z_0 = -\frac{\sqrt{2}}{2}(1 - i).$$

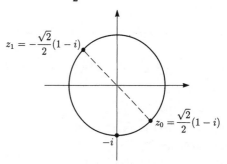

57

(ii) $4i = 4(\cos \pi/2 + i \sin \pi/2)$; the square roots of $4i$ are
$$z_k = 2(\cos(\pi/4 + k\pi) + i \sin(\pi/4 + k\pi)), \quad k = 0, 1.$$
Thus the principal square root is
$$z_0 = \sqrt{2} + \sqrt{2}i = \sqrt{2}(1 + i),$$
and the other root is
$$z_1 = -z_0 = -\sqrt{2}(1 + i).$$

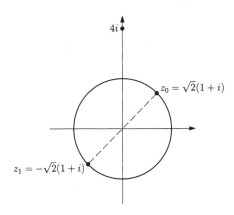

(b) (i) $-1 = \cos \pi + i \sin \pi$; the cube roots of -1 are
$$z_k = \cos\left(\frac{\pi}{3} + k\frac{2\pi}{3}\right) + i \sin\left(\frac{\pi}{3} + k\frac{2\pi}{3}\right), \quad k = 0, 1, 2.$$
The principal cube root is
$$z_0 = \cos \pi/3 + i \sin \pi/3 = \frac{1 + \sqrt{3}i}{2},$$
and the other roots are
$$z_1 = \cos \pi + i \sin \pi = -1,$$
$$z_2 = \cos 5\pi/3 + i \sin 5\pi/3 = \frac{1 - \sqrt{3}i}{2}.$$

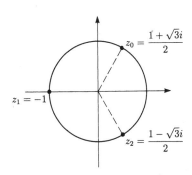

(ii) $-2 + 2i = 2\sqrt{2}(\cos 3\pi/4 + i \sin 3\pi/4)$; the cube roots of $-2 + 2i$ are
$$z_k = \sqrt{2}\left(\cos\left(\frac{\pi}{4} + k\frac{2\pi}{3}\right) + i \sin\left(\frac{\pi}{4} + k\frac{2\pi}{3}\right)\right),$$
$$k = 0, 1, 2.$$
The principal cube root is
$$z_0 = \sqrt{2}\left(\cos \pi/4 + i \sin \pi/4\right) = 1 + i,$$
and the other roots are
$$z_1 = \sqrt{2}(\cos 11\pi/12 + i \sin 11\pi/12),$$
$$z_2 = \sqrt{2}(\cos 19\pi/12 + i \sin 19\pi/12).$$

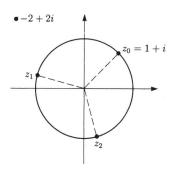

(c) (i) $\dfrac{\sqrt{2}}{2}(-1 - i) = \cos(-3\pi/4) + i \sin(-3\pi/4)$; the fourth roots of $\dfrac{\sqrt{2}}{2}(-1 - i)$ are
$$z_k = \cos\left(-\frac{3\pi}{16} + k\frac{\pi}{2}\right) + i \sin\left(-\frac{3\pi}{16} + k\frac{\pi}{2}\right),$$
$$k = 0, 1, 2, 3.$$
The principal fourth root is
$$z_0 = \cos(-3\pi/16) + i \sin(-3\pi/16),$$
and the other roots are
$$z_1 = \cos 5\pi/16 + i \sin 5\pi/16,$$
$$z_2 = \cos 13\pi/16 + i \sin 13\pi/16,$$
$$z_3 = \cos 21\pi/16 + i \sin 21\pi/16.$$

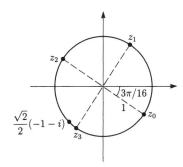

(ii) $-1 + i = \sqrt{2}(\cos 3\pi/4 + i \sin 3\pi/4)$; the fourth roots of $-1 + i$ are
$$z_k = 2^{1/8}\left(\cos\left(\frac{3\pi}{16} + k\frac{\pi}{2}\right) + i \sin\left(\frac{3\pi}{16} + k\frac{\pi}{2}\right)\right),$$
$$k = 0, 1, 2, 3.$$
The principal fourth root is
$$z_0 = 2^{1/8}(\cos 3\pi/16 + i \sin 3\pi/16),$$
and the other roots are
$$z_1 = 2^{1/8}(\cos 11\pi/16 + i \sin 11\pi/16),$$
$$z_2 = 2^{1/8}(\cos 19\pi/16 + i \sin 19\pi/16),$$
$$z_3 = 2^{1/8}(\cos 27\pi/16 + i \sin 27\pi/16).$$

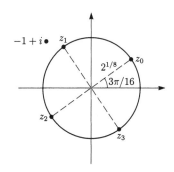

(d) (i) $-1 = \cos\pi + i\sin\pi$; the fifth roots of -1 are

$$z_k = \cos\left(\frac{\pi}{5} + k\frac{2\pi}{5}\right) + i\sin\left(\frac{\pi}{5} + k\frac{2\pi}{5}\right),$$
$$k = 0, 1, \ldots, 4.$$

The principal fifth root is

$$z_0 = \cos\pi/5 + i\sin\pi/5,$$

and the other roots are

$$z_1 = \cos 3\pi/5 + i\sin 3\pi/5,$$
$$z_2 = \cos\pi + i\sin\pi = -1,$$
$$z_3 = \cos 7\pi/5 + i\sin 7\pi/5,$$
$$z_4 = \cos 9\pi/5 + i\sin 9\pi/5.$$

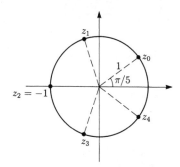

(ii) $-16 + 16\sqrt{3}i = 32(\cos 2\pi/3 + i\sin 2\pi/3)$; the fifth roots of $-16 + 16\sqrt{3}i$ are

$$z_k = 2\left(\cos\left(\frac{2\pi}{15} + k\frac{2\pi}{5}\right) + i\sin\left(\frac{2\pi}{15} + k\frac{2\pi}{5}\right)\right),$$
$$k = 0, 1, \ldots, 4.$$

The principal fifth root is

$$z_0 = 2(\cos 2\pi/15 + i\sin 2\pi/15),$$

and the other roots are

$$z_1 = 2(\cos 8\pi/15 + i\sin 8\pi/15),$$
$$z_2 = 2(\cos 14\pi/15 + i\sin 14\pi/15),$$
$$z_3 = 2(\cos 4\pi/3 + i\sin 4\pi/3),$$
$$z_4 = 2(\cos 26\pi/15 + i\sin 26\pi/15).$$

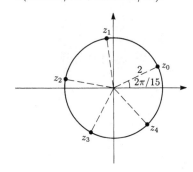

3.2 **(a)** Since $(x + iy)^2 = x^2 - y^2 + 2xyi$, equating the real parts and imaginary parts of

$$(x + iy)^2 = 3 + 4i,$$

we have

$$x^2 - y^2 = 3, \tag{1}$$
$$2xy = 4. \tag{2}$$

From Equation (2), $y = 2/x$ and substituting this in Equation (1), we obtain

$$x^2 - \frac{4}{x^2} = 3;$$

that is,

$$x^4 - 3x^2 - 4 = 0$$

or

$$(x^2 - 4)(x^2 + 1) = 0.$$

Since $x^2 \geq 0$, we have $x^2 = 4$, so that $x = \pm 2$; then $y = \pm 1$. The two solutions are therefore

$$x + iy = \pm(2 + i).$$

(b) Since $(x + iy)^2 = x^2 - y^2 + 2xyi$, equating the real parts and imaginary parts of

$$(x + iy)^2 = -5 + 12i,$$

we have

$$x^2 - y^2 = -5, \tag{3}$$
$$2xy = 12. \tag{4}$$

From Equation (4), $y = 6/x$ and substituting this in Equation (3), we obtain

$$x^2 - \frac{36}{x^2} = -5;$$

that is,

$$x^4 + 5x^2 - 36 = 0$$

or

$$(x^2 + 9)(x^2 - 4) = 0.$$

Since $x^2 \geq 0$, we have $x^2 = 4$, so that $x = \pm 2$; then $y = \pm 3$. The two solutions are therefore

$$x + iy = \pm(2 + 3i).$$

3.3 **(a)** Substituting $w = z^2$ in $z^4 - z^2 + 1 + i = 0$ gives

$$w^2 - w + 1 + i = 0,$$

which has solutions

$$w = \frac{1 \pm \sqrt{1 - 4(1 + i)}}{2}$$
$$= \frac{1}{2} \pm \frac{1}{2}\sqrt{-(3 + 4i)}$$
$$= \frac{1}{2} \pm \frac{i}{2}\sqrt{3 + 4i}.$$

From Exercise 3.2(a), $\sqrt{3 + 4i} = 2 + i$, and hence

$$w = \frac{1}{2} \pm \frac{i}{2}(2 + i)$$
$$= i, \ 1 - i.$$

(Of course, you may have spotted the factorization:

$$w^2 - w + 1 + i = (w - i)(w - 1 + i).)$$

Thus $z = \pm\sqrt{i}$ or $\pm\sqrt{1 - i}$. Since

$$i = \cos\pi/2 + i\sin\pi/2$$

and

$$1 - i = \sqrt{2}(\cos(-\pi/4) + i\sin(-\pi/4)),$$

we obtain the four solutions

$$z = \pm(\cos\pi/4 + i\sin\pi/4)$$

and

$$z = \pm 2^{1/4}(\cos(-\pi/8) + i\sin(-\pi/8)).$$

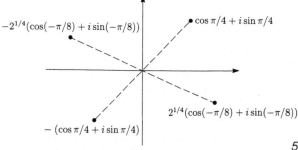

(b) Since the polynomial $z^3 - 4z^2 + 6z - 4$ has real coefficients and is of odd degree, we expect there to be at least one real solution. By trial and error (trying factors of 4), we find that $z = 2$ is a solution and hence

$$z^3 - 4z^2 + 6z - 4 = (z - 2)(z^2 - 2z + 2).$$

Since the solutions of $z^2 - 2z + 2 = 0$ are

$$z = \frac{2 \pm \sqrt{4 - 4 \times 2}}{2}$$
$$= \frac{2 \pm \sqrt{-4}}{2} = 1 \pm i,$$

the solutions to the original problem are

$$z = 2, \ 1 + i, \ 1 - i.$$

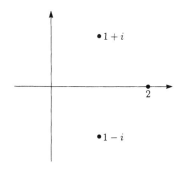

3.4 If z satisfies $p(z) = 0$, then

$$a_n z^n + \cdots + a_1 z + a_0 = 0.$$

Taking complex conjugates of both sides and using Theorem 1.1(b) gives

$$0 = \overline{a_n z^n + \cdots + a_1 z + a_0}$$
$$= \overline{a_n z^n} + \cdots + \overline{a_1 z} + \overline{a_0}$$
$$= a_n \overline{z}^n + \cdots + a_1 \overline{z} + a_0,$$

since $a_0, a_1, \ldots, a_n$ are real. Hence $p(\overline{z}) = 0$, as required.

Remark The solution to Exercise 3.3(b) provides an example of this result, whereas Exercise 3.3(a) shows that if the coefficients of a polynomial p are not all real, then solutions of $p(z) = 0$ need not occur in complex conjugate pairs.

Section 4

4.1 (a)

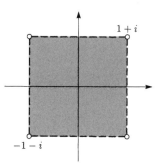

$$\{z : |\operatorname{Re} z| < 1, |\operatorname{Im} z| < 1\}$$

(b)

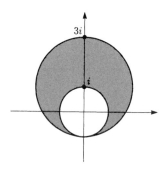

$$\{z : |z - i| \leq 2, |z| \geq 1\}$$

(c)

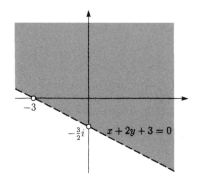

$$\{z : \operatorname{Re} z + 2 \operatorname{Im} z + 3 > 0\}$$

(d)

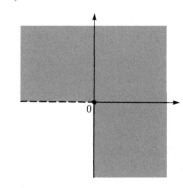

$$\{z : \operatorname{Re} z \geq 0\} \cup \{z : \operatorname{Im} z > 0\}$$

(e)

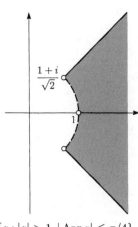

$\{z : |z| > 1, |\operatorname{Arg} z| \le \pi/4\}$

(f)

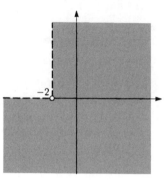

$\{z : -\pi < \operatorname{Arg}(z + 2)\}$
$\cap \{z : \operatorname{Arg}(z + 2) < \pi/2\}$

(g)

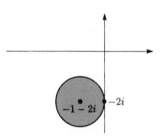

$\{z : |z + 1 + 2i| \le 1\}$

(h)

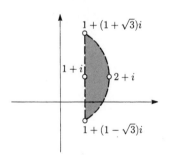

$\{z : \operatorname{Re} z > 1, |z - i| < 2\}$

(i)

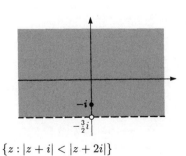

$\{z : |z + i| < |z + 2i|\}$

(j) First note that $z = 1$ is excluded from the set.

$$\operatorname{Re}\left(\frac{z+1}{z-1}\right) = \operatorname{Re}\left(\frac{(z+1)(\overline{z}-1)}{(z-1)(\overline{z}-1)}\right)$$
$$= \operatorname{Re}\left(\frac{z\overline{z} - z + \overline{z} - 1}{|z-1|^2}\right)$$
$$= \frac{\operatorname{Re}(|z|^2 - 2iy - 1)}{|z-1|^2}$$
$$= \frac{|z|^2 - 1}{|z-1|^2}.$$

So, for $z \ne 1$,
$$\operatorname{Re}\left(\frac{z+1}{z-1}\right) \le 0 \iff |z|^2 - 1 \le 0$$
$$\iff |z| \le 1.$$

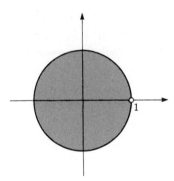

$\{z : \operatorname{Re}\left(\frac{z+1}{z-1}\right) \le 0\}$

(k)

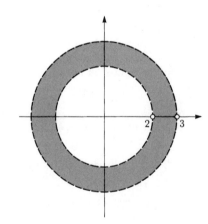

$\{z : |z| < 3\} - \{z : |z| \le 2\}$

(l) Since $z^2 + z - 2 = (z + 2)(z - 1)$,
$$\{z : z^2 + z - 2 = 0\} = \{1, -2\}.$$

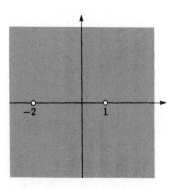

$\mathbb{C} - \{z : z^2 + z - 2 = 0\}$

Section 5

5.1 Throughout this solution, the appropriate version of the Triangle Inequality is given in parentheses.

(a) $|z + 3| \le |z| + |3|$ (usual form)
$$= 2 + 3 = 5, \quad \text{for } |z| = 2.$$

(b) $|z - 4i| \le |z| + |4i|$ (usual form)
$$= 2 + 4 = 6, \quad \text{for } |z| = 2.$$

(c) $|3z + 2| \le |3z| + |2|$ (usual form)
$$= 3|z| + 2$$
$$= 3 \times 2 + 2 = 8, \quad \text{for } |z| = 2.$$

(d) $|3z^2 - 5| \le |3z^2| + |5|$ (usual form)
$$= 3|z|^2 + 5$$
$$= 3 \times 2^2 + 5 = 17, \quad \text{for } |z| = 2.$$

(e) $|z^2 + z + 1| \le |z^2| + |z| + |1|$ (usual form)
$$= |z|^2 + |z| + 1$$
$$= 2^2 + 2 + 1 = 7, \quad \text{for } |z| = 2.$$

5.2 **(a)** $|z - 2| \ge ||z| - |2||$ (backwards form)
$$= |5 - 2| = 3, \quad \text{for } |z| = 5.$$

(b) $|z + 3i| \ge ||z| - |3i||$ (backwards form)
$$= |5 - 3| = 2, \quad \text{for } |z| = 5.$$

(c) $|z - 7| \ge ||z| - |7||$ (backwards form)
$$= |5 - 7| = 2, \quad \text{for } |z| = 5.$$

(d) $|2z - 7| \ge ||2z| - |7||$ (backwards form)
$$= |10 - 7| = 3, \quad \text{for } |z| = 5.$$

5.3 First we obtain upper and lower estimates for $|z^3 + 1|$, for $|z| = 4$, using appropriate versions of the Triangle Inequality (indicated in parentheses). Now
$$|z^3 + 1| \le |z^3| + 1 \quad \text{(usual form)}$$
$$= |z|^3 + 1$$
$$= 64 + 1 = 65, \quad \text{for } |z| = 4.$$
Also,
$$|z^3 + 1| \ge ||z^3| - |1|| \quad \text{(backwards form)}$$
$$= ||z|^3 - 1|$$
$$= |64 - 1| = 63, \quad \text{for } |z| = 4.$$

Next we obtain upper and lower estimates for $|z^3 - 1|$ for $|z| = 4$. Now
$$|z^3 - 1| \le |z^3| + |1| \quad \text{(usual form)}$$
$$= |z|^3 + 1$$
$$= 64 + 1 = 65, \quad \text{for } |z| = 4.$$
Also,
$$|z^3 - 1| \ge ||z^3| - |1|| \quad \text{(backwards form)}$$
$$= ||z|^3 - 1|$$
$$= |64 - 1| = 63, \quad \text{for } |z| = 4.$$
So
$$\left| \frac{z^3 + 1}{z^3 - 1} \right| = |z^3 + 1| \times \frac{1}{|z^3 - 1|}$$
$$\le 65 \times \frac{1}{63} = \frac{65}{63}, \quad \text{for } |z| = 4,$$
and
$$\left| \frac{z^3 + 1}{z^3 - 1} \right| = |z^3 + 1| \times \frac{1}{|z^3 - 1|}$$
$$\ge 63 \times \frac{1}{65} = \frac{63}{65}, \quad \text{for } |z| = 4.$$
Hence $m = 63/65$ and $M = 65/63$ will do.

5.4 If either z_1 or z_2 is zero, then the result evidently holds. For $z_1 \ne 0$ and $z_2 \ne 0$, let
$$z_1 = r_1(\cos\theta_1 + i\sin\theta_1)$$
and
$$z_2 = r_2(\cos\theta_2 + i\sin\theta_2).$$
Then
$$|z_1 + z_2|^2$$
$$= |(r_1\cos\theta_1 + r_2\cos\theta_2) + i(r_1\sin\theta_1 + r_2\sin\theta_2)|^2$$
$$= r_1^2 + r_2^2 + 2r_1 r_2(\cos\theta_1\cos\theta_2 + \sin\theta_1\sin\theta_2)$$
$$\qquad (\text{since } \sin^2\theta + \cos^2\theta = 1)$$
$$= r_1^2 + r_2^2 + 2r_1 r_2\cos(\theta_1 - \theta_2)$$
$$\le r_1^2 + r_2^2 + 2r_1 r_2 \qquad (\text{since } \cos\theta \le 1)$$
$$= (r_1 + r_2)^2$$
$$= (|z_1| + |z_2|)^2;$$
hence
$$|z_1 + z_2| \le |z_1| + |z_2|.$$